The IT Leader's Guide to SaaSOps

How to Secure Your SaaS Applications

VOLUME 2

by David Politis

PUBLISHED BY BETTERCLOUD

The IT Leader's Guide to SaaSOps (Volume 2):
How to Secure Your SaaS Applications is published by BetterCloud.

Author: David Politis | Editor: Christina Wang

https://www.bettercloud.com

Table of Contents

Introduction

Ten years ago, SaaS skeptics were urging the world not to move to the cloud.

The biggest issue holding IT professionals back was security. They were nervous about storing data in the cloud—on servers and systems they neither owned nor controlled. It wasn't safe, they thought, to trust third parties with their data (or entire business infrastructure for that matter).

The naysayers were vocal. Richard Stallman, founder of the Free Software Foundation and creator of the computer operating system GNU, told the world in 2008 that using web-based apps like Gmail was "worse than stupidity . . . if you use a proprietary program or somebody else's web server, you're defenseless. You're putty in the hands of whoever developed that software."[1]

Those were some pretty strong words, but he wasn't alone. Technology journalist Jack Schofield wrote something similar that year in *The Guardian*, saying: "Look, if you have data online, you can lose access to it at any second, through hacking, an idle whim, a simple mistake, or some financial or even natural disaster. In fact, calling it 'the cloud' is a good metaphor, because it's insubstantial and easily blown away."[2]

> "In fact, calling it 'the cloud' is a good metaphor, because it's insubstantial and easily blown away."
>
> Jack Schofield, 2008

But fast-forward a decade, and we've done a complete 180.

"IT Pros Agree: Security Is Better in the Cloud,"[3] proclaimed one 2017 InfoWorld headline.

1 Johnson, Bobbie. "Cloud Computing Is a Trap, Warns GNU Founder." *The Guardian*, Guardian News and Media, 29 Sept. 2008, www.theguardian.com/technology/2008/sep/29/cloud.computing.richard.stallman.
2 Schofield, Jack. "When Google Owns You.... Your Data Is in the Cloud." *The Guardian*, Guardian News and Media, 6 Aug. 2008, www.theguardian.com/technology/blog/2008/aug/06/whengoogleownsyouyourdata.
3 Linthicum, David. "IT Pros Agree: Security Is Better in the Cloud." *InfoWorld*, InfoWorld, 31 Mar. 2017, www.infoworld.com/article/3185757/it-pros-agree-security-is-better-in-the-cloud.html.

"Why Email Is Safer in Office 365 Than on Your Exchange Server,"[4] declared another 2017 headline from CIO.com.

Today, the situation has reversed entirely. Now IT teams are nervous about storing data on-prem. It was safer, they realized, to trust cloud providers than use legacy systems. Better to put it on a server or system you *don't* own or control.

Why did IT do an about-face?

Because SaaS app providers are better at securing your applications than you ever will be. They eat, sleep, and breathe network security. It's their core competency, not yours. Their reputation and entire business model depend on it. They have bigger economies of scale, more technical expertise, state-of-the-art data centers, and dedicated teams of world-class security experts, allowing them to secure environments in ways most companies could never hope to match (or afford to match!). Their platforms are faster and more reliable. They have failover solutions, uptime guarantees, and disaster recovery testing.

That said, SaaS platforms only secure their applications. *You* must secure how they're used.

That said, SaaS platforms only secure their applications. *You* must secure how they're used.

But doing that—especially in a way that doesn't sacrifice user productivity—is a tricky undertaking.

This book tackles that quandary. It's a deep dive into a new category, SaaSOps, and a new concept: *user interactions*. Your users are the new perimeter in the digital workplace, and securing their interactions is the key to securing your SaaS data. This book is meant to get you thinking critically about who and what your users interact with; how to classify,

4 Branscombe, Mary. "Why Email Is Safer in Office 365 than on Your Exchange Server." *CIO*, CIO, 7 Mar. 2017, www.cio.com/article/3177718/why-email-is-safer-in-office-365-than-on-your-exchange-server.html.

evaluate, and analyze these interactions; how to think about what trust means to you; and how to secure interactions in a way that aligns with your security philosophy.

Note: With its detailed focus on data protection, this book is heavily focused on the security side of SaaSOps. For a more macro, high-level framework on the management side of SaaSOps, I encourage you to read my first book. You can find *The IT Leader's Guide to SaaSOps (Volume 1): A Six-Part Framework for Managing Your SaaS Applications* on Amazon. These books will provide a complementary look at SaaSOps, and I recommend reading them in tandem.

CHAPTER ONE

What Is SaaSOps?

CHAPTER ONE HIGHLIGHTS

What Is SaaSOps?

SaaSOps is all about automating the operational administrative and security tasks necessary to keep an organization's SaaS applications running effectively.

A direct response to new challenges created by SaaS, **SaaSOps focuses on managing and securing** *how* people use SaaS apps.

SaaSOps is based on two core principles: **management and security**.

When it comes to protecting SaaS data, **IT teams are best suited for the job**. Nobody understands the complexities or the business impact of SaaS applications like IT does.

SaaS Operations, or SaaSOps, is a new discipline for IT and security teams.

It's called a few different names in the wild: digital workplace ops, IT operations, SaaS administration, cloud office management, end user computing. But at the heart of it, these terms all essentially refer to the same thing: all the new IT responsibilities, technologies, processes, and skills needed to successfully enable your org through SaaS apps.

What SaaSOps is

In the same way that DevOps was born out of a need for faster deployment (with fewer defects), SaaSOps was born out of a need for faster, more efficient management of SaaS apps (with less human error).

In the same way that DevSecOps bakes security into development workflows (to protect against new risks that CI/CD introduces), SaaSOps bakes security into IT workflows (to protect against new risks that SaaS introduces).

SaaSOps is all about automating the operational administrative and security tasks necessary to keep an organization's SaaS applications running effectively.

In the same way that DevOps creates lasting competitive advantages, SaaSOps enables the business to achieve strategic goals more quickly and securely.

Here's my definition of SaaSOps:

SaaSOps *noun*
a philosophy referring to how software-as-a-service (SaaS) application data is managed and secured through centralized and automated operations (Ops).

SaaSOps is all about automating the operational administrative and security tasks necessary to keep an organization's SaaS applications running effectively. It's for managing and securing *how* people use SaaS apps. In doing so, it strives to extract business value from SaaS.

SaaSOps is based on two core principles:

- **Management:** SaaSOps makes sure the right users have access to the right data at the right time and automates historically manual work like onboarding, offboarding, and user lifecycle management. By automating these tasks, IT can focus on strategic work that drives the business forward, thereby becoming a partner to the business, not a cost center.
- **Security:** SaaSOps protects mission-critical data in SaaS apps so that companies avoid data breaches or leakage, hefty compliance fines, loss of IP, loss of competitive advantage, and/or business disruption.

Both of these principles rely heavily on centralization. Because SaaS apps are siloed, the only way to manage and secure them effectively is to centralize all the data first. You need to have a complete, consolidated view of all the data objects connected across apps. Only then can you start automating workflows and protecting SaaS data.

Now, SaaSOps also encompasses several secondary areas. These include things like spend management, data backup and recovery, end user training, business workflow automation, integrations, and shadow IT discovery. They all play a part in operating SaaS apps effectively and holistically.

But management and security are the most critical components of SaaSOps. Without these two principles in place, enabling, scaling, and protecting the business in question is impossible.

What SaaSOps isn't

To be clear, SaaSOps does not refer to the uptime, performance, or availability of the SaaS application; those are all the SaaS vendor's responsibility. Or, in the case of hosted private cloud/public cloud custom applications, these aspects are to be addressed by using an Application Performance Management (APM) vendor for the underlying cloud infrastructure.

Nor does SaaSOps refer to application infrastructure security. Again, that's handled by the SaaS vendors. They take care of the physical security of the data centers, securing the storage of data and the hardware and software that underlie the infrastructure.

Instead, SaaSOps focuses on what's in IT's wheelhouse: the operational tasks related to managing users and the ways those users engage with SaaS applications.

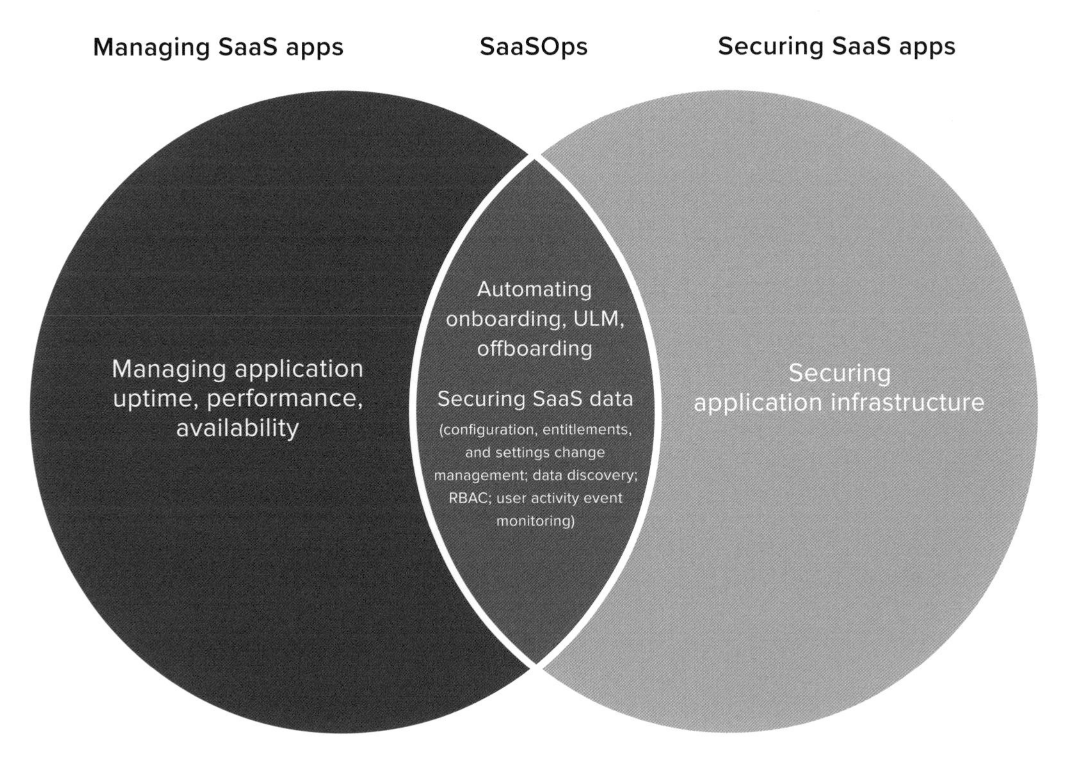

Figure 1: SaaSOps sits at the intersection of SaaS application management and security. It focuses on how *people use SaaS apps.*

Why is SaaSOps emerging now?

SaaSOps is emerging now because IT's role is changing.

IT's mandate has always been about enabling business through technology. They've long helped employees do what they need to do to generate revenue. But that decree has evolved. It now encompasses security as well, because that's becoming a bigger priority for IT. For one, CIOs are often blamed for data breaches; they're the "sacrificial lambs." Think of the high-profile data breaches at Target and Equifax. Heads rolled, and it was the CIOs who took the fall.

Additionally, when it comes to protecting SaaS application data, IT teams are best suited for the job. They understand all the nuances and complexities of these systems. They're the ones who bear responsibility for SaaS apps end to end: deploying, managing, integrating, and training end users to use them. Nobody understands the specifics of these systems or their business impact like IT does. They're empowering the entire workforce with SaaS every day. They understand, better than anybody in the org, where this mission-critical data is stored—and why it's so hard to protect it.

SaaSOps is a direct response to new challenges created by SaaS, which include:

- The morass of manual, repetitive tasks required of IT
- Alert fatigue
- Privilege creep
- Shadow IT
- Data sprawl (SaaS sprawl)
- Configuration drift
- Increased freedom for end users to share data and collaborate, resulting in a new type of insider threat
- Lack of visibility into risky settings and entitlements
- Siloed information contained within individual SaaS apps
- A shifting perimeter

These challenges exist because SaaS is like the Wild West; it's become a chaotic "free-for-all" frontier. Previously, enterprise application installation on corporate servers required IT's help, but today anyone can buy, deploy, and use SaaS. It gives users escalating freedom and control, which is often left unchecked. Missteps in configurations and entitlements run amok.

SaaS also creates a massive, complex, interconnected data sprawl that grows by the day. Think of all the data objects that reference, interact, control, and/or rely on each other: users, groups, mailboxes, files, folders, records, contacts, calendars, third-party apps, logs, metadata, permissions, devices, etc. How can IT begin to make sense of this sprawl?

Despite these IT challenges, employees today expect—and even demand—to use SaaS apps. SaaS is no longer something that can be avoided or resisted.

The goal of SaaSOps

The goal of SaaSOps is to regain control over this Wild West. A key part of SaaSOps is managing SaaS applications more effectively. It aims to give IT more visibility into their murky SaaS environment, more actionable insights and less noise, and a way to automate routine operational tasks while embedding security best practices.

On a similar note, securing SaaS data is the other key part of SaaSOps. It aims to protect organizations against data exposure, data theft, and risky user activity.

Together, these two SaaSOps components enable IT to gain control over this new frontier. Achieving that control starts with defining acceptable use policies for SaaS apps and then executing on those policies. SaaSOps emphasizes the ability to create, enforce, and optimize policies for mission-critical SaaS apps.

SaaSOps emphasizes the ability to create, enforce, and optimize policies for mission-critical SaaS apps.

Benefits of SaaSOps

What's the upside?

SaaSOps enables productivity and reduces friction, empowering employees to work faster and more efficiently with SaaS. Benefits include time savings, cost savings, reductions in human error, simplified compliance processes, data security, reduction in IT ticket volume, and a more productive and engaged workforce. Automating operational processes ultimately allows IT to focus on strategic initiatives, empowering IT to be a partner to the business.

In the following chapters, we'll go in-depth into SaaSOps, the digital workplace, user interactions, and what you can do today to start securing them.

CHAPTER TWO

The Digital Workplace

(And Why the User Is the Center of It)

CHAPTER TWO HIGHLIGHTS

The Digital Workplace (And Why the User Is the Center of It)

SaaSOps introduces a new concept for IT in the digital workplace: **Control and secure the user**, not the infrastructure or perimeter.

Your **users are closest to your data**, which is what you're ultimately trying to protect.

Authentication vs. authorization. Authentication is the process of granting access to apps by verifying that users are who they claim to be. Authorization is what comes next. It grants access to specific SaaS data, configurations, resources, or functions.

For companies moving to SaaS, authentication solves the first order problem: **identity and access**. Authorization solves the second order problem: user interactions.

Once upon a time, tech stacks were homogeneous.

One single vendor used to power a company's infrastructure and enterprise application needs. For example, Microsoft could meet all of an organization's email, chat, storage, and video conferencing needs through Exchange, Communicator, and SharePoint.

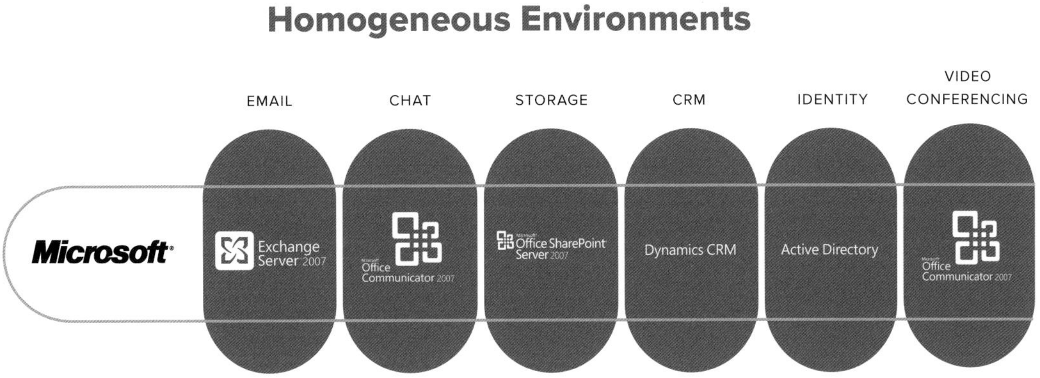

Homogeneous Environments

EMAIL
CHAT
STORAGE
CRM
IDENTITY
VIDEO CONFERENCING
Microsoft
Exchange Server 2007
Office Communicator 2007
Office SharePoint Server 2007
Dynamics CRM
Active Directory
Office Communicator 2007

Figure 2: In a perfect world, IT departments could meet all their organization's needs through a single vendor.

But those days are long gone. As more specialized solutions enter the market and technology becomes easier to access, SaaS applications are compounding infrastructure complexity. This is giving rise to heterogeneous environments, where an organization's business needs are fulfilled by a wide assortment of best-of-breed cloud apps from multiple vendors.

Heterogeneous Environments

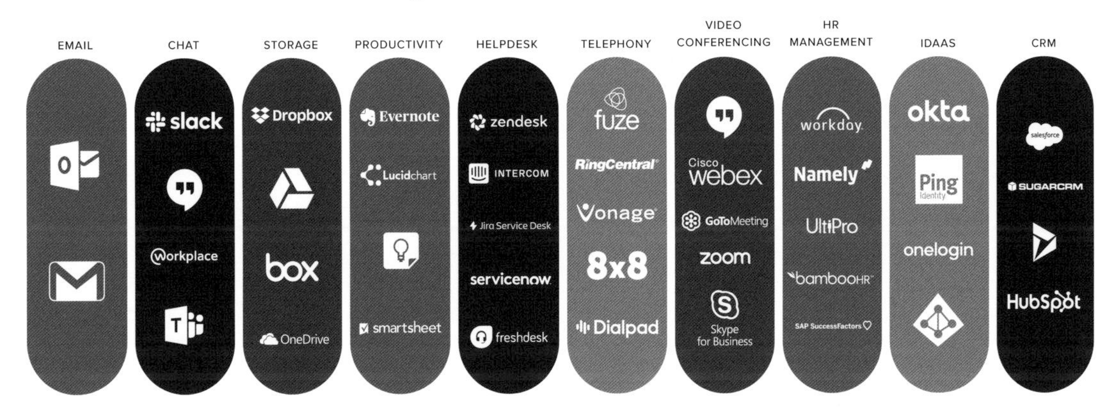

Figure 3: In a heterogeneous environment, organizations are free to pick and choose their cloud applications.

A modern workforce might use Gmail for email but Slack for chat, Zendesk for ticketing, Zoom for video conferencing, Workday for an HRIS, and Salesforce for a CRM. This is what I think of as the *digital workplace*.

People have been talking about the digital workplace for years, but search interest in the term has soared recently:

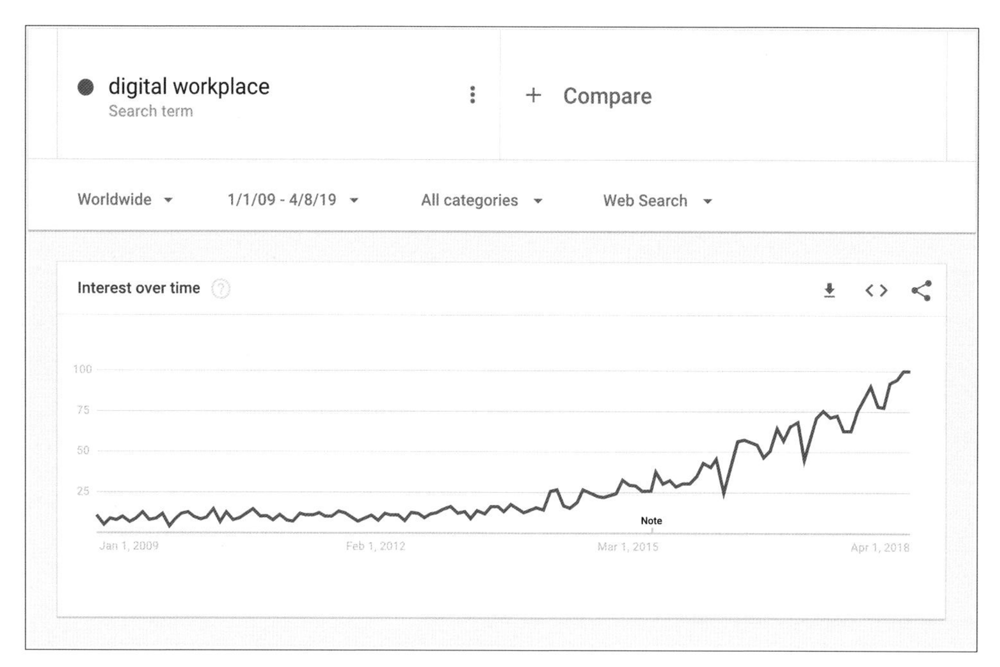

Figure 4: Interest in the search term "digital workplace" has increased over time from 2009 to 2019. (Source: Google Trends)

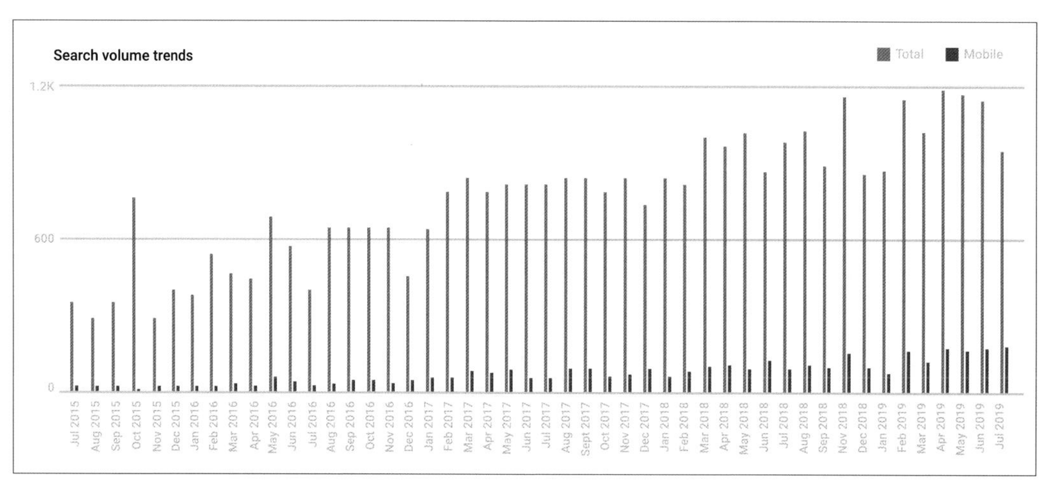

Figure 5: Total search volume for "digital workplace" has increased over time from July 2015 to June 2019. (Source: Google Keyword Planner)

As employees use digital technology to do their jobs and as software and devices are increasingly connected, it's not surprising that interest in the term "digital workplace" has risen over time. There are dozens of definitions out there, but this is mine:

digital workplace ***noun***

a professional environment where employees are enabled and empowered to use the latest technology to stay engaged and productive.

Heterogeneous environments make the digital workplace possible. Companies can pick and choose the best possible software for every job, making it easier than ever to get work done. When the experience of work is delivered through apps, tools, and business systems in the cloud, and when your tech stack is driving productivity, a digital workplace is born.

In the digital workplace, users have come to expect these best-of-breed apps. According to a 2018 report by Forrester Research, the majority of workers consider the ability to access files and apps from any location or device and the ability to collaborate with colleagues in real time on documents and files to be "very important" or "critical" capabilities.[5] In fact, 77 percent of employees say they prefer technologies that give them the freedom to choose how and where to get their work done.[6]

Employees are aware of all the benefits SaaS applications offer. With the consumerization of enterprise software, they're using these apps both personally and professionally. They've used them at previous companies, and they're demanding collaborative, flexible environments wherever they go.

> If you don't (or can't) enable your employees with the latest technology, at best you'll have an unproductive, unengaged workforce, while at worst you'll have a talent shortage.

As a result, IT teams must support this new way of working. Tasked with providing the best user experience, they need to empower

5 *Rethink Technology In The Age Of The Cloud Worker: A Spotlight On The Employee*. Forrester Research, July 2018, cloud.google.com/files/chrome-enterprise/helpcenter/misc/rethink_technology_in_the_age_of_the_cloud_worker_employee_spotlight_forrester_tlp.pdf.
6 Ibid.

the workforce with SaaS. They need to enable innovation to grow the business. If you don't (or can't) enable your employees with the latest technology, at best you'll have an unproductive, unengaged workforce, while at worst you'll have a talent shortage.

IT has no choice. They must find a way to enable productivity and strategic goals while also keeping up with operational demands and ensuring critical data is protected. To do that, you have to start by controlling and securing the user.

In the digital workplace, the user is the centerpiece of it all

SaaSOps introduces a new concept for IT in the digital workplace: Control and secure the user, not the infrastructure or perimeter.

Perimeter-based security is obsolete. Devices and data are not hosted on-premises anymore, so the process of creating a network in a corporate data center and protecting it with a firewall is no longer useful or relevant. The shift to the cloud means the network perimeter has dissolved. With the advent of SaaS, employees no longer just work nine 'til five at the office on one device. They use a panoply of unmanaged devices—smartphones, mobile devices, tablets, Chromebooks—whenever they want from multiple locations outside the LAN. This makes the traditional perimeter an abstraction, something that no longer exists in practice.

All of these macro changes mean that IT must shift to a different security model. In SaaSOps, the area of focus is something new: your user, and how they're using SaaS apps. Why?

- **Your users are closest to your data, which is what you're ultimately trying to protect.** Today, your confidential business data, trade secrets, IP, employee data, and customer data all lives in your SaaS apps. Your users are interacting with this data every day to do their jobs—changing, updating, and sharing it continuously. To protect your data, you have to control and secure who has access to it.

To protect your data, you have to control and secure who has access to it.

- **Your users have a lot of freedom and power within SaaS apps (and, as a result, IT teams are losing control).** Users can share data freely with just about anyone inside or outside the org: colleagues, partners, customers, contractors, even competitors. They can adjust permissions and sharing settings on their own, add themselves to distribution lists and groups, and share data publicly on the web. Of course, all of this freedom is by design. It's what makes SaaS such a powerful productivity tool. But the very beauty of SaaS—the openness and ease of sharing data—is also its most dangerous risk.

- **Your users are creating a tremendous SaaS sprawl.** They're using multiple apps (e.g., Office 365, G Suite, Slack, Dropbox, third-party apps, custom apps) on multiple devices (e.g., laptops, Chromebooks, tablets, mobile phones). As SaaS adoption grows, so does the amount of data living in those SaaS apps, which in turn creates an enormous information sprawl. The bigger that sprawl is, the harder it is to get visibility and stay in compliance.

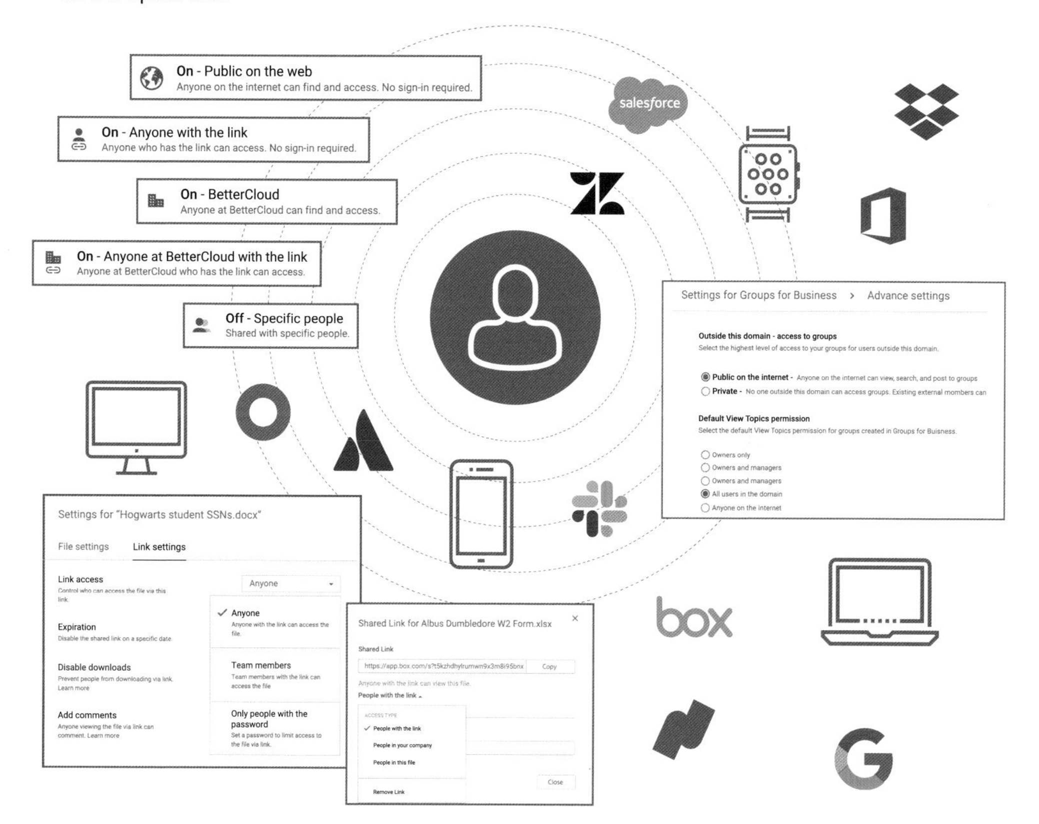

Figure 6: The user is the centerpiece of the digital workplace. They use dozens of SaaS apps across multiple devices and create a massive SaaS sprawl.

We've reached a tipping point. Because the traditional network perimeter has vanished, a new perimeter has emerged: the user.

But this new frontier creates a looming question: How do you control and secure that user?

We've reached a tipping point. Because the traditional network perimeter has vanished, a new perimeter has emerged: the user.

Authentication is only half the battle

First, let's make an important distinction: authentication vs. authorization. Both play a part in controlling and securing the user. Authentication is the process of granting access to apps by verifying that users are who they claim to be. Authorization is what comes next. It grants access to specific SaaS data, configurations, resources, or functions.

For years now, cloud security strategies have incorporated identity and access management (IAM) and, more recently, identity-as-a-service (IDaaS) tools. One of the most important benefits of SaaS is that users can access apps from any device at any time from anywhere. For this to happen, there has to be a secure authentication (i.e., connection) between that user and their apps. Before granting access to corporate data, companies have to verify: Are you really who you say you are?

Authentication is the process of granting access to apps by verifying that users are who they claim to be. Authorization is what comes next.

With tools like multi-factor authentication (MFA), privileged identity management (PIM), single sign-on (SSO), privileged access management (PAM), and directory extensions, identity-based security has tightened the authentication protocols in place. It ensures that only authorized people from authorized locations can access authorized resources. By centrally managing user identities and controlling their access to resources, cloud-based IAM helps secure this new perimeter.

However, while authentication is a good way to begin controlling and securing users, it's only half the battle. It's akin to a doorman at the entrance of a building: They grant entry,

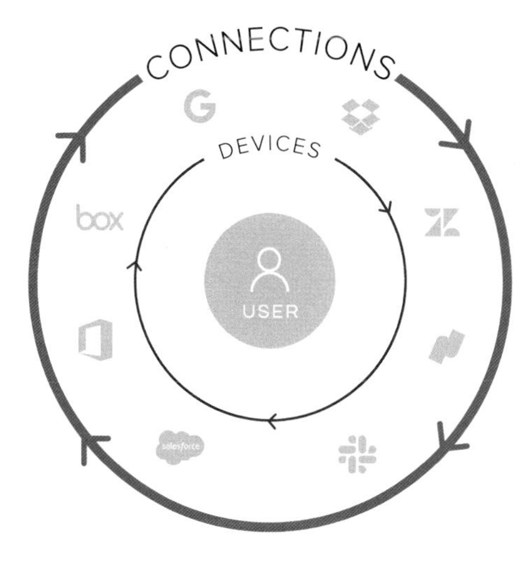

Figure 7: The various layers of a digital workplace. Your users (the centerpiece of the digital workplace) need devices and secure connections to do their jobs.

but what happens after that? Security doesn't stop there. Inside the building, there are CCTV surveillance cameras, alarm systems, and additional guards, all to monitor what people are doing, look for abnormal activity, and make sure they continue to behave in a safe manner after they enter.

Similarly, once a SaaS app authenticates a user, security shouldn't stop there. It's critical that you understand what's actually happening within those applications *post-authentication* too.

This is the authorization layer. What data are users authorized to access, download, forward, copy, export, change, add, update, preview, edit, print, upload, share, delete, create, and so on? What entitlements or settings can users modify? What groups and distribution lists can they join, view, edit, delete, or share? That's where the concept of user interactions come in.

Put another way, authentication solves the first order problem for companies moving to SaaS: identity and access. Authorization solves the second order problem: user interactions.

CHAPTER THREE

User Interactions:

How Work Gets Done in Your SaaS Apps

CHAPTER THREE HIGHLIGHTS

User Interactions: How Work Gets Done in Your SaaS Apps

Data protection in SaaS environments starts with **securing user interactions**.

Interactions are the actions your users are taking to get work done: the processes they're performing inside SaaS apps, the people they're interacting with, and the data they're interacting with.

Securing interactions is critical in SaaSOps because: 1) SaaS is the system of record, 2) SaaS makes it easy to expose (or steal) data, as can already be seen in the news, and 3) the biggest security threat is from negligent end users.

There are **three types of interactions**: interactions with your own data, interactions with trusted users, and interactions with untrusted users.

The definition of a trusted user can vary across orgs. It really depends on how you and your organization define *trust*—specifically, who you feel comfortable trusting and what kind of data you feel comfortable trusting them with.

Interactions create an enormous data sprawl. The larger the sprawl, the higher the risk of human error or negligence, even though users may have the best of intentions.

It's not sufficient to know that your users are authorized to use their SaaS apps. It's a solid step in your cloud strategy, but holistically securing your SaaS data requires more work. Next, you also need to know what's happening *inside* the apps after users gain access. This is what I call a "user interaction."

Interactions are the actions your users are taking to get work done—the processes they're performing inside SaaS apps, the people they're interacting with, and the data they're interacting with.

Interactions are the actions your users are taking to get work done—the processes they're performing inside SaaS apps, the people they're interacting with, and the data they're interacting with. Essentially, it's everything that comes after authentication, after the doorman grants you entry. It's *how* they're using SaaS apps.

My definition of an interaction is:

interaction *noun*
a user's activity in relation to other users, data, and devices.

Work in the digital workplace is being done through interactions today. Yes, you use devices and SaaS apps to get your job done, but devices are merely access points—a way to get on the internet (and, by extension, to SaaS apps). The actual work—the activity you engage in to achieve a result—occurs through interactions *within* those apps.

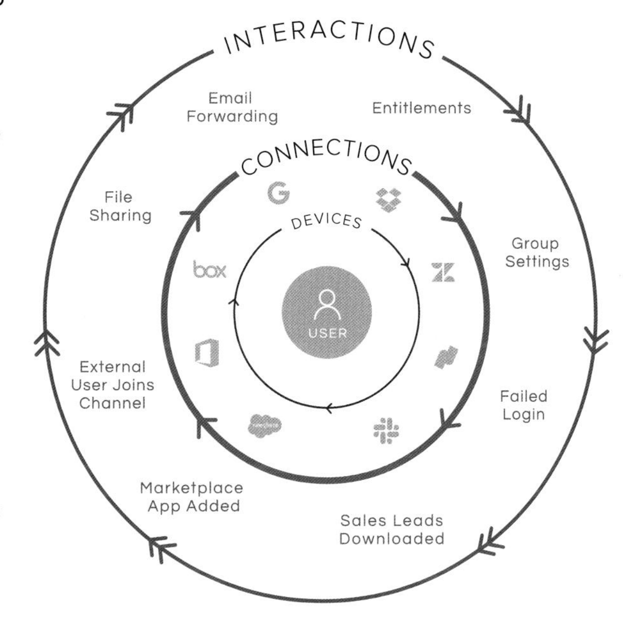

Figure 8: The various layers of a digital workplace, including interactions. This includes actions like file sharing, email forwarding, and entitlements.

Examples of interactions: who and what you interact with

Here are some examples (some riskier than others) of how people interact in the digital workplace:

Interactions with users:

- You send a Drive file to an intern in Slack.
- You share a PDF in SharePoint Online with a part-time designer.
- You share a spreadsheet in Dropbox with a competitor.
- You email a confidential financial report through Outlook to an employee in the sales department.
- You share a roadmap in Google Drive with a former colleague.
- You assign an Office 365 license to a new employee.
- You share a SoW in Box with a partner.
- You invite an external copywriter to join the product launch team in Microsoft Teams.
- You make a Helpdesk Admin a Super Admin in G Suite.

Interactions with data objects:

- **Files**
  - You export 200 reports (.csv) from Salesforce in one day.
  - You download physician's treatment logs (.docx) and photos (.jpg) from Dropbox.
  - You upload an executable file (.exe) to OneDrive.
  - You delete a Google Drive folder containing assembly line schematics.
  - You export an employee compensation summary report (.pdf) from Workday.
  - You email an RFP (.pdf) to an agency using Outlook.
  - You open a spreadsheet (.xlsx) containing employee bank account numbers and Social Security numbers.
- **Folders.** You download an entire folder of contracts from SharePoint Online.
- **Calendars.** You share an editorial calendar in Google Calendar with the sales team.

- **Slack channels.** Your PR contractor, a Single-Channel Guest in Slack, joins the #pr channel.
- **Groups.** You create a new public Office 365 group that anyone in the company can view and join.
- **Permissions.** You edit dashboard permissions in Splunk.
- **Tickets.** You submit an incident report in ServiceNow that then gets assigned to a support agent.
- **Emails.** You create a filter in Gmail that automatically forwards any work email containing the words "sales leads" to your personal Gmail account.
- **Surveys.** You create a customer survey in SurveyMonkey that has a public link.
- **Meetings.** You join a Zoom meeting with a customer.
- **Trello boards.** You keep track of feature requests on a public Trello board.
- **GitHub repositories.** You upload and commit an existing file to a GitHub repository.
- **Applications**
    - You add 30 apps to your company app catalog in OneLogin for SSO.
    - You create your own custom app to streamline resource allocation.
    - You download the Pokémon Go app (a third-party app) using your work credentials.

Interactions with devices:

- **Laptops**. You wipe a MacBook Air using Jamf.
- **Mobile devices.** You automatically install security policies and Wi-Fi settings on corporate-owned iOS devices using AirWatch.
- **Interactive whiteboards.** You sketch and co-edit a flowchart in real-time using a Cisco Webex Board.
- **Echo devices.** You sit in a conference room and say, "Alexa, join my meeting." Alexa finds the upcoming meeting on your calendar, turns on the video conferencing equipment, and connects you to the meeting.
- **VOIP phones.** After a call ends, Voice AI automatically detects and displays action items or next steps mentioned during the call.

Why securing interactions is critical in SaaSOps

SaaS applications are a double-edged sword. They empower us to collaborate and communicate at scale, which is what makes SaaS so powerful, but at the same time, these very interactions can introduce new liabilities and formidable challenges into your environment.

As a result, SaaSOps calls for new operational controls and processes to protect SaaS data. To do that, you must secure interactions because:

- **SaaS is the system of record.** Today, organizations are trusting SaaS vendors to house mission-critical, irreplaceable data. When you're storing sensitive information like PII, trade secrets, sales pipeline data, employee salary information, and tax documents in SaaS apps that employees are interacting with on a daily basis, it's especially important to secure this activity.

- **SaaS makes it easy to expose (or steal) data, as can already be seen in the news.** SaaS apps have garnered publicity thanks to interactions like the ones discussed in the previous section. Misconfigured Google Groups, Box files, Google Calendars, Trello boards, and GitLab instances, as well as innocuous SaaS app integrations, have resulted in data exposure incidents. Here are just a few headlines:

UN EXPOSES SENSITIVE DATA ON PUBLIC TRELLO BOARDS[7]
TechTarget, 2018

Dozens of Companies Leaked Sensitive Data Thanks to Misconfigured Box Accounts[8]
TechCrunch, 2019

Samsung Spilled SmartThings App Source Code and Secret Keys[9]
TechCrunch, 2019

Hundreds of Companies Expose PII, Private Emails Through Google Groups Error[10]
ZDNet, 2017

18F CAUSED A "DATA BREACH" USING SLACK[11]
Nextgov, 2016

Doxed by Microsoft's Docs.com: Users Unwittingly Shared Sensitive Docs Publicly[12]
Ars Technica, 2017

Corporate Data Slips Out Via Google Calendar[13]
PCWorld, 2007

These incidents were accidental, but intentional actions like trade secret theft have also made headlines. For example, Zynga sued two ex-employees in 2016, alleging that one of them had downloaded 10 Google Drive folders and taken over 14,000 files and approximately 26 GB of extremely sensitive, highly confidential Zynga information to a competitor.[14] Similarly, in the Uber vs. Waymo trade secrets trial of 2018, a Google security engineer testified that an ex-employee had downloaded 14,107 files (9.74 GB) and exported several confidential and proprietary documents from Google Drive to a personal device before leaving the company.[15]

- **SaaS applications are creating a new generation of insider threats, where the biggest risk is from well-meaning but negligent end users.** When it comes to insider threats, the biggest threat is not from splashy saboteurs seeking revenge; 62 percent of IT professionals believe that the biggest security threat actually comes from well-meaning but negligent end users.[16] These are your ordinary employees. They are particularly dangerous because they have access to critical assets but lack the training or knowledge to keep sensitive information safe as they do their jobs. They may not understand the consequences of their interactions. For companies

> Sixty-two percent of IT professionals believe that the biggest security threat actually comes from well-meaning but negligent end users.

7 Bacon, Madelyn. "UN Exposes Sensitive Data on Public Trello Boards." *SearchSecurity*, 28 Sept. 2018, searchsecurity.techtarget.com/news/252449650/UN-exposes-sensitive-data-on-public-Trello-board
8 Whittaker, Zack. "Dozens of Companies Leaked Sensitive Data Thanks to Misconfigured Box Accounts – TechCrunch." *TechCrunch*, 11 Mar. 2019, techcrunch.com/2019/03/11/data-leak-box-accounts/
9 Whittaker, Zack. "Samsung Spilled SmartThings App Source Code and Secret Keys – TechCrunch." *TechCrunch*, 8 May 2019, techcrunch.com/2019/05/08/samsung-source-code-leak
10 Osborne, Charlie. "Hundreds of Companies Expose PII, Private Emails through Google Groups Error." *ZDNet*, 27 July 2017, www.zdnet.com/article/simple-settings-failure-in-google-groups-caused-exposure-of-private-company-employee-data/.
11 Ravindranath, Mohana. "Watchdog: 18F Caused a 'Data Breach' Using Slack." *Nextgov.com*, 28 Nov. 2017, www.nextgov.com/cio-briefing/2016/05/watchdog-18f-caused-data-breach-using-slack/128288
12 Gallagher, Sean. "Doxed by Microsoft's Docs.com: Users Unwittingly Shared Sensitive Docs Publicly." *Ars Technica*, 27 Mar. 2017, arstechnica.com/information-technology/2017/03/doxed-by-microsofts-docs-com-users-unwittingly-shared-sensitive-docs-publicly/
13 McMillan, Robert. "Corporate Data Slips Out Via Google Calendar." PCWorld, 17 Apr. 2007, www.pcworld.com/article/130868/article.html
14 Farivar, Cyrus. "Zynga Case." *DocumentCloud*, www.documentcloud.org/documents/3227518-Zynga-Case.html
15 Panzarino, Matthew. "Declaration of Gary Brown (PUBLIC)." *DocumentCloud*, www.documentcloud.org/documents/3515476-Declaration-of-Gary-Brown-PUBLIC.html.
16 "State of Insider Threats in the Digital Workplace 2019." *BetterCloud Monitor*, 29 Apr. 2019, www.bettercloud.com/monitor/insider-threats-digital-workplace-2019

that are powered by SaaS apps, the negligent end user has even more freedom to unintentionally expose sensitive information. SaaS gives users extensive control and power, but to err is human. It's extraordinarily easy to make a simple misconfiguration mistake, especially when presented with dozens of complex sharing settings across multiple apps.

That said, many IT professionals today are not securing user interactions effectively. On a recent webinar poll, we found out that 90 percent of IT professionals either felt their system to secure user interactions and activity within SaaS apps was insufficient or they didn't have a system altogether.[17]

SaaS data is not easy to secure. Specifically, 75 percent of IT professionals believe that cloud storage/file sharing and email are the biggest security challenges.[18] Tackling SaaS security is difficult because SaaS is relatively new; not enough time has passed for official certifications or industry best practices to exist. There is no foundational level of knowledge yet, no ITIL for SaaS. In fact, 78 percent of IT professionals are just getting started managing SaaS apps or are still teaching themselves.[19]

Seventy-five percent of IT professionals believe that cloud storage/file sharing and email are the biggest security challenges.

If you haven't grappled with these challenges or their precursors already, you will at some point in your SaaS journey—it's inevitable. But by going in-depth into interactions—what they mean for IT, how to think about them, how to secure them—this book is meant to give you a head start on protecting your SaaS data. Hopefully, it'll help you avoid common SaaSOps pitfalls.

Let's start by building foundational knowledge. A good first step is to make sure you understand the three various types of interactions, as well as how the concept of trust fits into them.

17 Buyers, Maddie. "Closing the Gap in SaaS Security." *BetterCloud Monitor*, 28 Mar. 2019, www.bettercloud.com/monitor/closing-the-gap-in-saas-security
18 "State of Insider Threats in the Digital Workplace 2019." *BetterCloud Monitor*, 29 Apr. 2019, www.bettercloud.com/monitor/insider-threats-digital-workplace-2019/
19 Wang, Christina. "The Top Security Blind Spots in Your SaaS Environment." *BetterCloud Monitor*, 14 May 2018, www.bettercloud.com/monitor/the-top-security-blind-spots-recap/

The three types of interactions

Your users have hundreds, if not thousands, of interactions on any given day to get work done. In aggregate, your company likely has tens of millions of interactions. Broadly speaking, there are three categories of interactions:

1) Interactions with your own data

One type of interaction is between you and your own data. Now, to be clear, it's not really "your" data. You don't own it; the company does. But for example, this type of data might mean your own email or your Chrome extensions that you installed. It might mean a private file in Box that you haven't shared with anyone else and that only you can access.

2) Interactions with trusted users

Another type of interaction is between you and a trusted user. Trusted users are people who *should* have the access that they do. Their access is sanctioned.

Trusted users have authorized access to the right data at the right time. But who, exactly, qualifies as a trusted user? That's not so cut and dried. Defining a trusted user is a nuanced challenge because it really depends on your company.

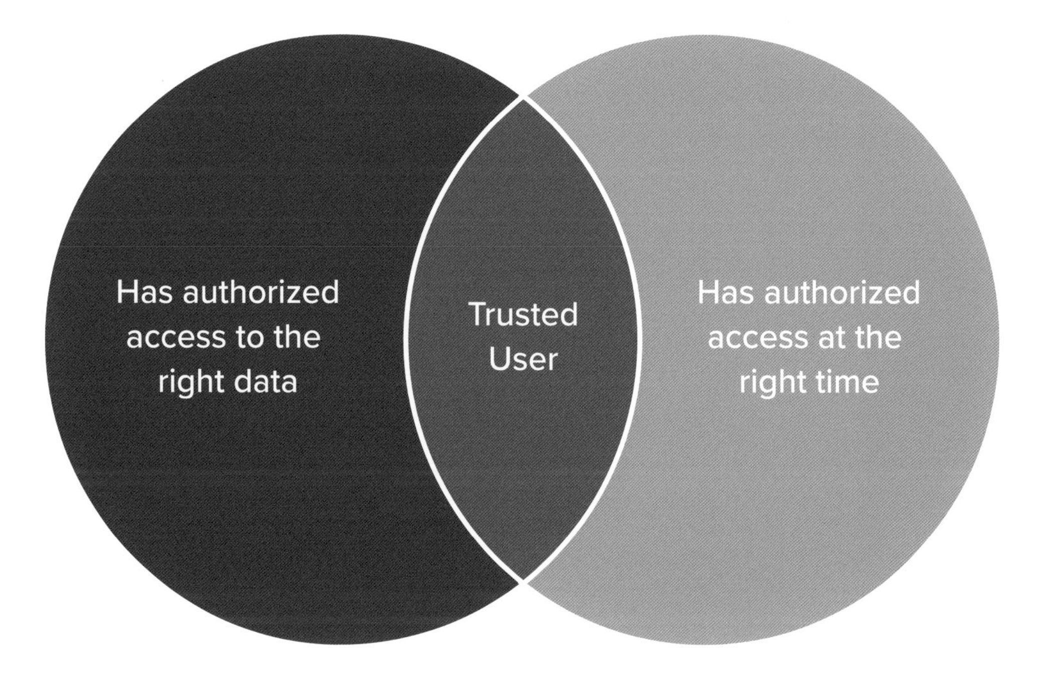

Figure 9: A trusted user has authorized access to the right data at the right time.

Let's take a look at this statement:

Everyone inside my org is a trusted user.

For some companies, this statement may be true. Others may disagree.

For example, if an engineer has access to confidential financial data that they shouldn't be privy to, would you still consider them a trusted user? Or if two departments in a financial institution were supposed to be divided by a "Chinese Wall" but could access each other's data, would you still consider them trusted users? Or what if all employees were able to read the messages in the HR distribution list? Would you still consider them trusted users?

Now let's take a look at the inverse of that statement:

Everyone outside my org is an untrusted user.

Again, for some companies, this statement may be true. You might not want your employees to collaborate with anyone outside your company. But other companies may disagree. If your employees are collaborating with external users like partners, resellers, contractors, clients, or board members on projects, then you might consider certain people outside of your organization to be trusted users.

Here's where it gets murky: Let's say your services team shares a SoW with a partner. On the surface, that partner is a trusted user. But what if that SoW contains confidential pricing information that the partner shouldn't be seeing? Would you still consider them a trusted user?

For example, if an engineer has access to confidential financial data that they shouldn't be privy to, would you still consider them a trusted user?

Or take another example: Your marketing manager invites an outside PR contractor to join the #marketing Slack channel. At first glance, the contractor is a trusted user. But who else is in that Slack channel? What else is being discussed, and what kinds of documents are being shared in it? Do you still consider them a trusted user? What about when their contract ends? That's where the line gets blurry.

In these examples, there is no right or wrong answer. It really depends on how you and your organization define *trust*—specifically, who you feel comfortable trusting and what kind of data you feel comfortable trusting them with.

3) Interactions with untrusted users

Classifying untrusted users is much more straightforward. An untrusted user is someone who unequivocally *should not* have access to your data.

Some examples of interactions with untrusted users include:

- A strategy exec shares proprietary research with a competitor
- An HR manager shares a file containing employee Social Security numbers and salaries with an ex-employee
- A junior IT admin is given super admin rights across all SaaS apps

Untrusted users are a security risk. Interactions like these can result in unauthorized access to sensitive data, excess admin privileges, compliance violations and fines, data breaches, loss of IP or trade secrets, loss of revenue, negative PR, loss of consumer trust, and more.

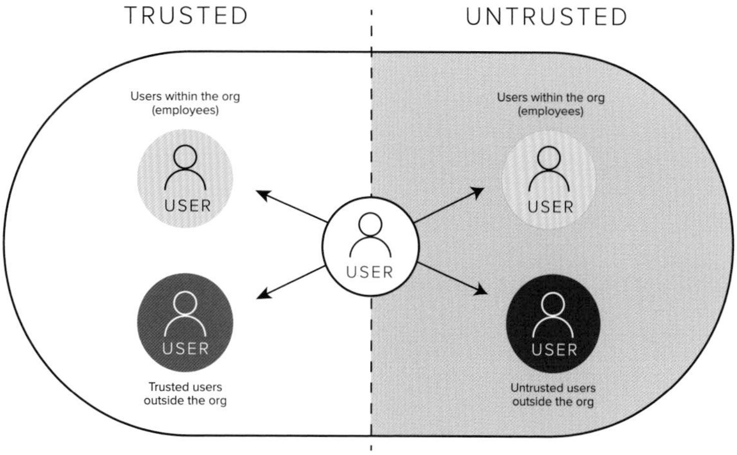

Figure 10: Users are having interactions with trusted and untrusted users both inside and outside the org.

The sprawl of user interactions

Because SaaS applications were expressly designed to foster collaboration, sharing data is simple. But due to this openness and the simplicity of sharing, interactions are like the digital version of kudzu: a sprawling, massive web. They spread in all directions throughout your environment and grow out of control quickly. Left unchecked, they're nearly impossible to manage.

To visualize this, here are a few examples of how easily interactions can sprawl out of control.

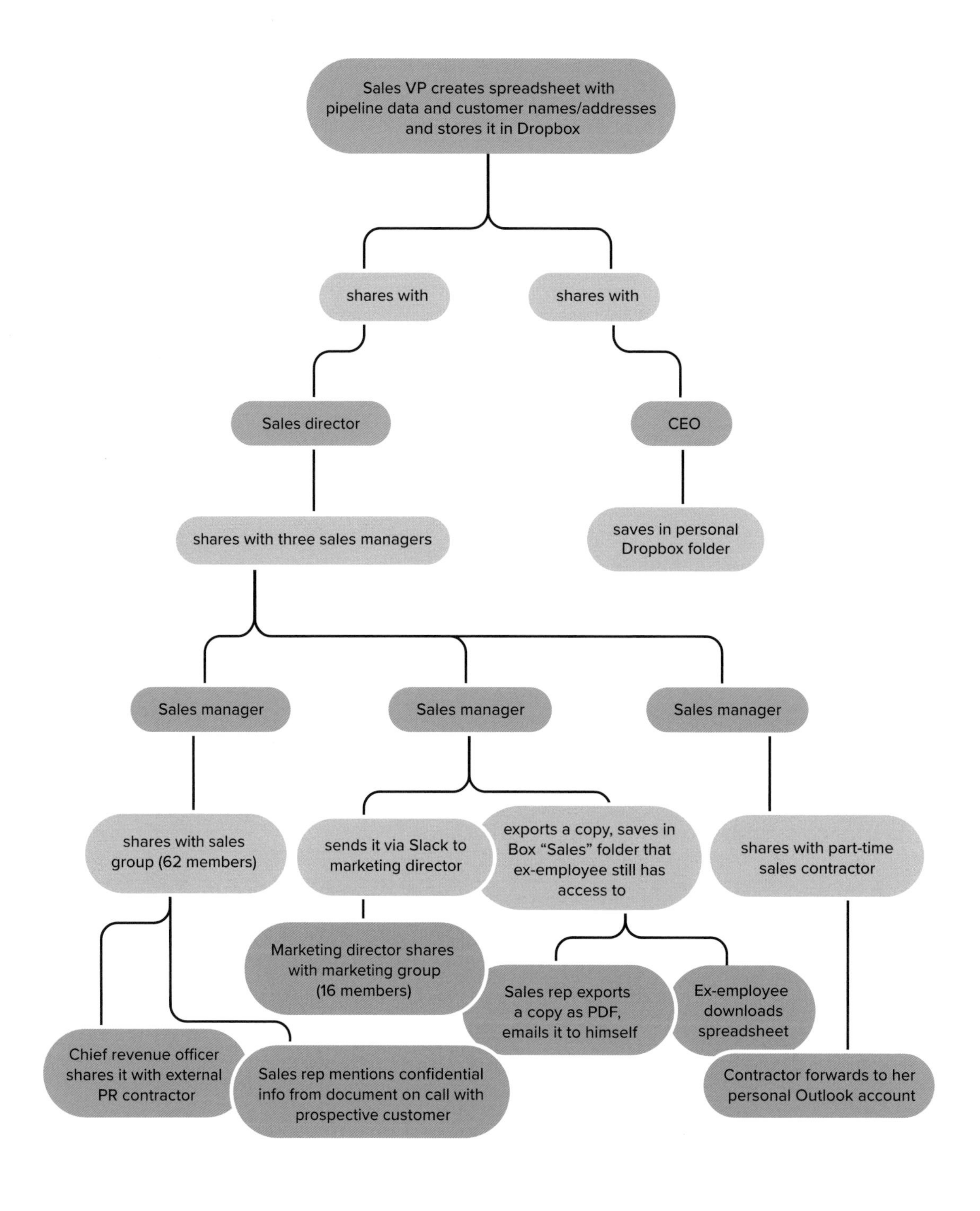

Figure 11: A visualization of how the sprawl of interactions can spiral out of control when it comes to file sharing.

FILES

Let's take files as an example. A sales VP may store a spreadsheet in Dropbox containing confidential sales pipeline data and customer names and addresses. Initially, his interactions are innocuous. He only intends for his CEO and sales director to access this data, so he shares it with the two of them.

But from there, the sales director shares it with her three sales managers, who in turn download, export, forward, email, save, and share it with internal (and external) users, who in turn do the same. The more people the file is shared with, the bigger the sprawl is. The growth becomes exponential. The larger the sprawl, the higher the risk of human error or negligence, even though users may have the best of intentions. Essentially, the sprawl is never-ending. IT has little visibility into this sprawl, but it continues to grow exponentially across your org as people do their jobs and collaborate.

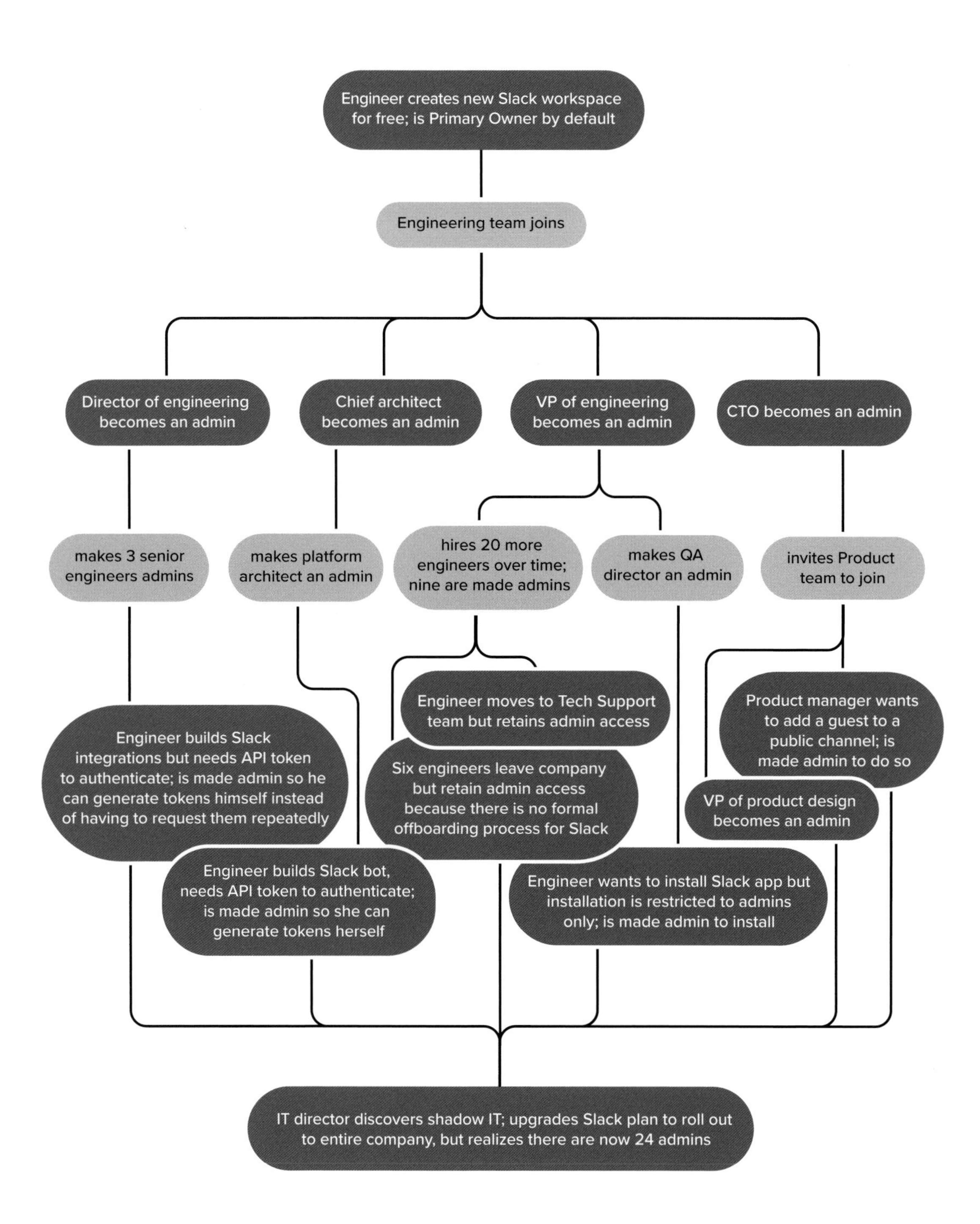

Figure 12: A visualization of how over-assignment of admin privileges can happen over time.

EXCESSIVE ADMIN PRIVILEGES

Here's another example focusing on excessive admin rights. Based on a true story, this example shows how easily admin privileges can be over-assigned.

In a shadow IT world, it's not unusual to see an individual team (e.g., Engineering) adopting Slack without IT's approval or knowledge.

Initially, there might only be a few admins. But eventually, that number swells. Admins beget admins: Managers might invite others and make them admins too.

In this example, as engineers build Slack integrations and bots, they need API tokens to authenticate them. It's easier to give them all admin access so they can generate tokens on their own, rather than having to continuously request them. Similarly, if they want to install a Slack app but installation is restricted to admins only, it's easier to just give them admin access. Convenience trumps security.

Unfortunately, this means admin rights are often handed out like candy, with no oversight. Privilege creep sets in too. People switch roles but retain admin access. Additionally, if a Slack workspace starts out as shadow IT, there is often no formal offboarding process. Nobody is officially in charge of deactivating accounts. As a result, ex-employees can easily retain Slack access—and admin access at that.

This story happened to one of our customers. When we were onboarding them, we helped them discover a whopping 79 admins (most of them engineers) in their Slack instance, which started out as shadow IT. They had no idea. With our help, they managed to reduce the number of admins to two.

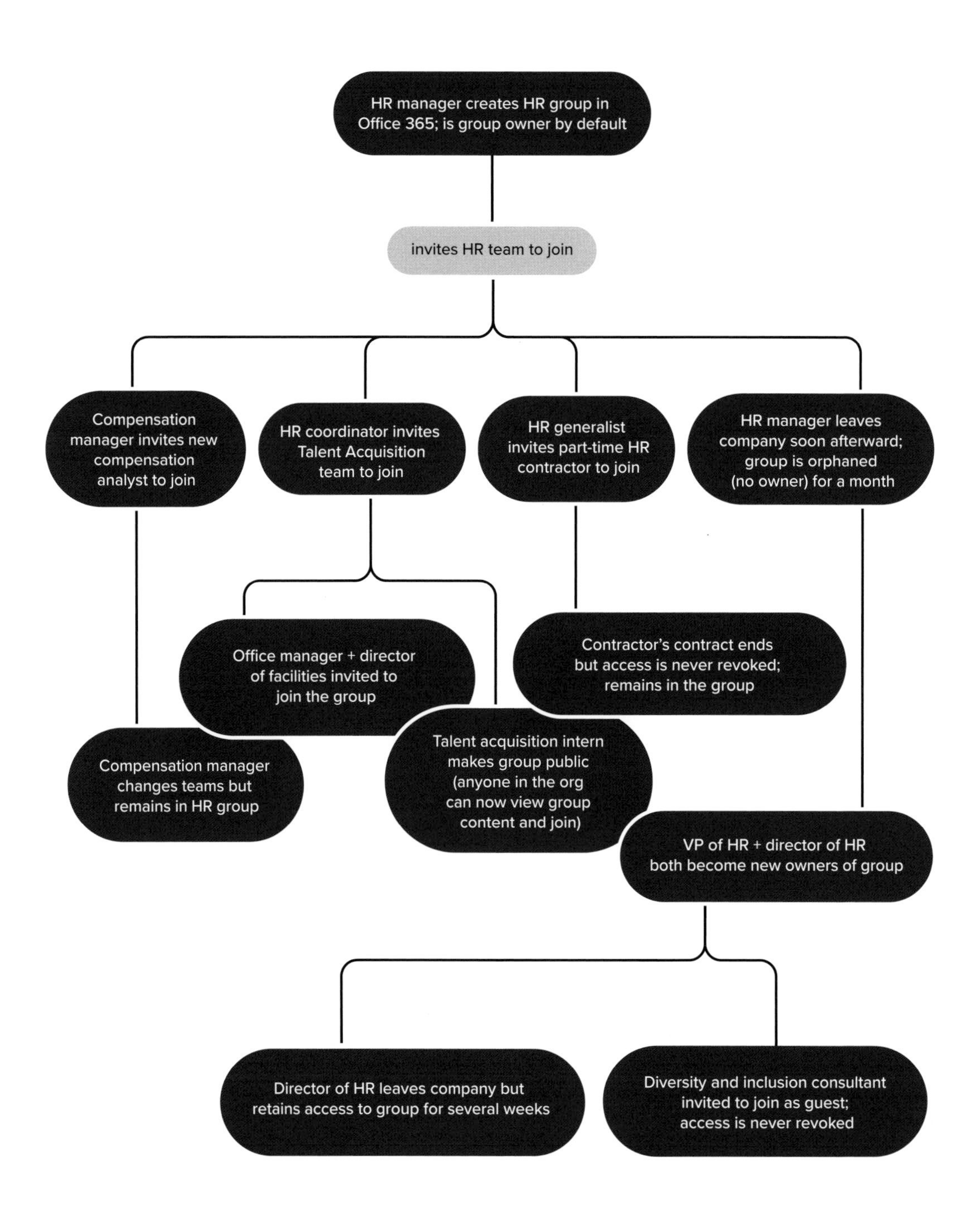

Figure 13: A visualization of "interaction sprawl" caused by groups over time.

GROUPS

The last example of "interaction sprawl" has to do with groups. Over time, group ownership can become messy, and users can play fast and loose with group privacy settings. If the group contains sensitive information, this can become a security issue.

For example, let's say the director of HR creates an HR group. The default setting for groups in Office 365 is "Open," meaning users can create their own groups as needed without having to bother IT. He invites his HR team to join. Over time, the group grows. Members add other members—both internal employees and external guests. The group owner changes. Part-time contractors join, but their access is never revoked. Privacy settings are unintentionally changed from private to public, allowing anybody in the org to view the group's content and join.

These are just three examples, but interactions create a never-ending sprawl that spreads to just about every crevice in the digital workplace. So long as employees use SaaS applications to collaborate and get work done, the sprawl will continue growing.

Interactions by department

Up to this point, I've focused on why it's critical to secure interactions across all your core business apps like Slack, G Suite, or Office 365. But think about each department in your organization. Each team also has interactions within their own department-specific SaaS apps.

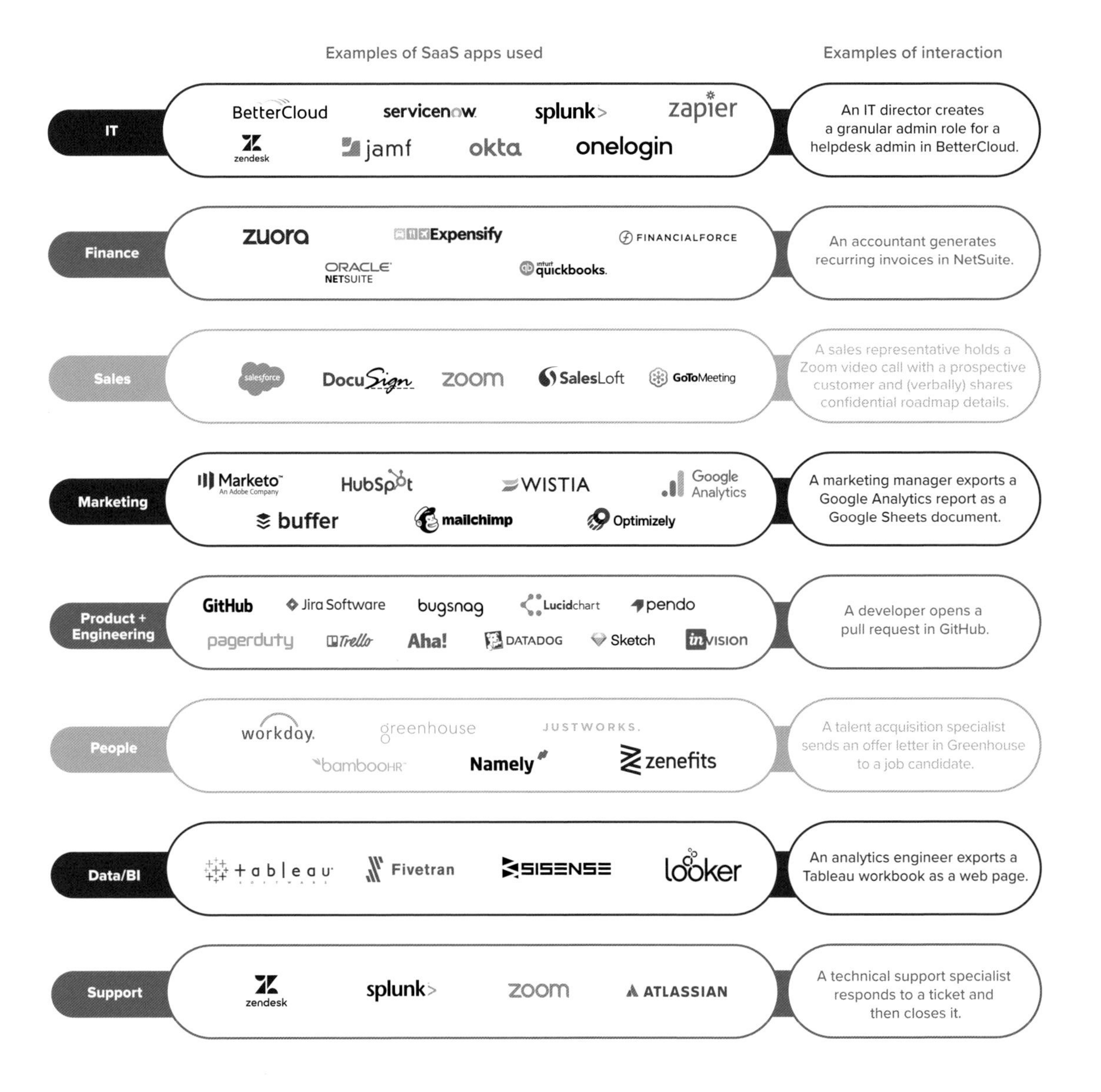

Figure 14: An example of how employees use department-specific SaaS apps and the types of interactions they're having within them.

For just about every business use case, there is now a SaaS app. And, because SaaS is the system of record now, these apps all house sensitive customer data, employee data, code, etc. for each of your departments. It's important to remember that interactions extend beyond just those found in your core SaaS apps; they also permeate your entire organization from team to team through department-specific apps.

There are multiple types of interactions in the digital workplace. Each one has its own nuances. Imagine how you would make the correct security decision for each interaction—and then multiply that task by millions. To do that in a legacy security center, how many employees would you need to manage these millions of interactions? How do you accomplish that in a SaaS environment? In this messy, complex sprawl of interactions, you need a systematic approach to determine what is a secure and authorized user interaction and what is not.

To better understand how to do that, let's take a look at what I call interaction *dimensions*.

CHAPTER FOUR

How to Classify Interactions in Your SaaS Apps

CHAPTER FOUR HIGHLIGHTS

How to Classify Interactions in Your SaaS Apps

Not all interactions are created equal.

How you evaluate interactions depends on key factors like **identity** (who's doing the interacting), **time** (when and how often an interaction occurs), and **data classification** (how sensitive the data is).

No single factor can define how risky an interaction is. Your definition of trust will vary based on the complexity of your IT environment, industry and compliance requirements, and business needs.

In the world of SaaSOps, not all interactions are created equal.

Similar to the idea of trusted vs. untrusted users, some interactions are far riskier or more important than others. Some warrant more scrutiny; others, less.

But how do you classify interactions? What implications do they have for IT teams and your data protection strategy? That all depends on several key factors I call "dimensions."

Identity: Who's doing the interacting?

Identity—*who* is doing the activity, and *who* they're sharing data with—plays a critical role in how you think about interactions. You can think about identity in a few ways:

- **Business role.** Is the user a C-level executive or is he an intern? Senior management will have access to more sensitive and valuable data than, say, a part-time intern. It's why "whaling"—phishing attacks that specifically target executives—exists. "[T]he victim is considered to be high-value, and the stolen information will be more valuable than what a regular employee may offer. The account credentials belonging to a CEO will open more doors than an entry-level employee,"[20] writes *CSO Online*'s Fahmida Y. Rashid.

- **External vs. internal.** Does the user belong to your organization or is he outside it? External users (like contractors and partners) need special attention since they are outside the company. While they may be trusted users in the sense that you want to collaborate with them, the least privilege model dictates that their access be restricted in some way, whether by:

  - Time (e.g., they have a two-month contract, so their access must expire at some point)

  - Groups (e.g., they only need access to one channel in Slack, so they're a Single-Channel Guest)

  - Rights (e.g., they only have Viewer rights on a Dropbox folder, which prevents them from editing folder contents or inviting new members)

20 Rashid, Fahmida Y. "Types of Phishing Attacks and How to Identify Them." *CSO Online*, CSO, 27 Oct. 2017, www.csoonline.com/article/3234716/types-of-phishing-attacks-and-how-to-identify-them.html.

- **Part-time vs. full-time.** Is the user a part-time (temporary) or full-time worker? According to a Forrester research report, temporary employees have a much higher turnover rate than full-time employees; they are much less loyal.[21] On top of that, many enterprises do not have dedicated security policies and controls for temporary workers due to IT staff capacity limitations or the misguided belief that short-term workers "don't have enough time" to be dangerous.[22] It's common to let formal training and IT protocol slip through the cracks when it comes to temporary workers. They're often brought on during peak times so there's little time for proper protocol, and their employment is fleeting. As with external workers, temporary workers should have access only to the amount of information required to do their jobs, and there should be appropriate controls in place.

According to a Forrester research report, temporary employees have a much higher turnover rate than full-time employees; they are much less loyal.

- **Job longevity.** How long has the user been employed with the company? A well-tenured employee will have access to—and generated—much more data than a relatively new employee. He may also have more loyalty to the company than, say, a brand new employee who just started a week ago.

- **Remote vs. in-office.** Does the user work remotely or are they in the office? Remote workers pose their own set of challenges—the major risks being the inability to enforce security, a lack of commitment to security best practices, and risky behavior on their part.[23] One of the most common mistakes remote workers often commit, especially when they work from home, is sharing their company-issued laptop with friends and family.[24] Additionally, accessing unsafe networks is a major issue. If a remote worker is using public Wi-Fi (say, at a coffee shop) to access company data, that free hotspot doesn't require any authentication to establish a network connection. This means

21 Cser, Andras, et al. "Forrester." *Identity And Access Management Mitigates Risks During Economic Uncertainty*, 26 Jan. 2009, www.forrester.com/report/Identity And Access Management Mitigates Risks During Economic Uncertainty/-/E-RES46572#.
22 "'Five Golden Rules' For Reducing Security Risk Posed By Temporary Holiday Workers." *Dark Reading*, 17 Dec. 2009, www.darkreading.com/risk/five-golden-rules-for-reducing-security-risk-posed-by-temporary-holiday-workers/d/d-id/1132636.
23 "Cyber Security & Remote Working: What Are the Risks, What Can You Do?" *Comparitech*, 28 May 2019, www.comparitech.com/blog/information-security/security-remote-working/.
24 Fahey, Ryan. "Security Awareness Issues for Remote Workers." *Infosec Resources*, resources.infosecinstitute.com/category/enterprise/securityawareness/security-awareness-roles/security-awareness-issues-for-remote-workers/.

that hackers can potentially access all the info a remote worker sends on the internet, including account logins, credit card information, confidential business data, you name it.

- **Trusted vs. untrusted.** Of course, the definition of trust will vary across orgs. Based on who the user is and what kind of data they have access to, are they trusted or untrusted in your eyes?

The digital workplace introduces new types of employees. It enables collaboration with people both inside and outside your org. While this shift is a game-changer for productivity, it also opens up a new set of liabilities, considerations, and implications. *Who* a user is will affect what kinds of data they have access to, how much data they have access to, how long they have access, their work habits, etc. All of these identity-related factors can help you think about what a trusted (or untrusted) interaction means to you and how strictly you want to secure them.

Time: When are these interactions happening?

Thinking about time—not only *when*, but *how often* an interaction occurs—can help you decide if it's to be trusted or untrusted. Think about:

- **Time of day.** Are these interactions happening during work hours or outside work hours? Of course, interactions during normal work hours are to be expected. Sporadic interactions taking place outside normal work hours might not raise any eyebrows either. After all, SaaS enables us to work at any time. For many companies, the work-life boundary is blurred. But if your work culture is one where employees typically "disconnect" from all things work-related after five p.m., after-hours interactions may signal an issue. How you interpret time of day depends on your organizational culture.

- **Time of week.** Are these interactions happening on weekdays or the weekend? This question is similar to the one above. The digital workplace means work and personal time are more fluid, but weekend interactions are generally less common.

- **Time of year.** Are these interactions happening on holidays or non-holidays? If you work in a seasonal industry, is there a flurry of interactions during the off-season?

- **Vacation/sick days.** Are interactions happening while a user is on vacation, sick, or otherwise out of the office?

- **Frequency.** Are interactions happening at an expected frequency? What's the normal baseline frequency for your org? If there's an abnormally high number of interactions in a short timeframe, this can be a major indicator that something is awry. Here are two examples:
    - Multiple, consecutive failed logins in a short window. Perhaps the user has forgotten their password, or perhaps they are the victim of a password-guessing brute force attack.
    - An unusually high number of file downloads in a short window. According to *CSO Online*, modifying large numbers of files in a short period of time can be a warning sign of an insider threat.[25] An exiting employee may be tempted to download files in bulk and take them to a new company.

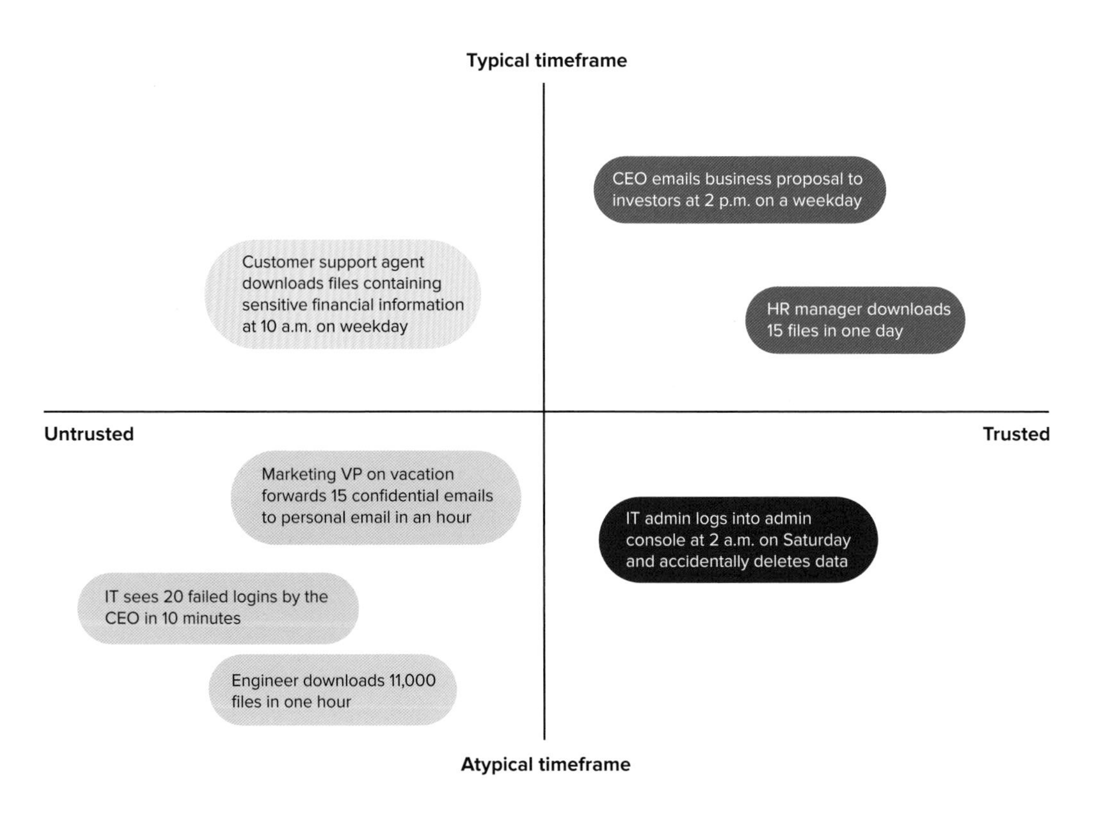

Figure 15: A perceptual map depicting how interactions can vary by time and trust.

25 Swinhoe, Dan. "7 Warning Signs of an Insider Threat." *CSO Online*, CSO, 3 Dec. 2018, www.csoonline.com/article/3323402/7-warning-signs-of-an-insider-threat.html.

Data classification: How sensitive is the data?

Users interact with data every day to do their jobs, but not all data is created equal.

How sensitive (or non-sensitive) is the data accessed by the interaction? If bank account numbers were exposed publicly, you'd likely be very concerned; a community calendar, maybe not so much.

"Sensitive" can mean a few different things. Think about:

- **Personal information.** Are users interacting with PII? Examples of PII include biometric data; passport numbers; Social Security numbers; bank account numbers; driver's license numbers; and medical, educational, financial, and employment information. Under many privacy and data protection laws, photographs are often regulated as "personal data" too. But even with photographs, some are more sensitive than others. For example, you might respond differently to the public exposure of photos of adults vs. photos of children. (This happened to a daycare center who became one of our customers. We helped them discover that they had accidentally exposed 15,000 children's photos publicly.)
- **Business information.** Are users interacting with any confidential information that, if discovered by the general public or a competitor, could pose a risk? Examples include trade secrets, patents, product roadmaps, sales pipeline data, credit card numbers, financial information, customer lists, exit interview notes, tax documents, M&A plans, proprietary research, employee salary information, etc.

Examples include trade secrets, patents, product roadmaps, sales pipeline data, credit card numbers, financial information, customer lists, exit interview notes, tax documents, M&A plans, proprietary research, employee salary information, etc.

- **Dangerous or harmful information.** Are employees using profanity in their files or phone conversations? For EDUs, are students referencing drugs, bullying, suicide, or self-harm in their interactions? It's important to remember that sensitive information isn't limited only to personal or business data. I've seen this firsthand with our customers, who have detected thousands of dangerous interactions unrelated to PII or IP that were equally as critical. Here are two of the most memorable examples that come to mind:

- One of our EDU customers, a school with 1,500 employees, regularly scanned their environment for keywords related to self-harm and bullying. One day, they discovered a suicide note written by a female student. The IT team immediately notified administrators and guidance counselors, who were able to provide crisis counseling and avert a tragedy. Without the ability to detect sensitive data, the school (and student's family) might never have known about this.

- This might be the most unsettling customer story I'll ever hear. One of our customers, a manufacturing company with nearly 30,000 employees, made an alarming discovery one day. They discovered that an employee in Malaysia was using her company Google+ account to share and spread ISIS propaganda and other terrorist-related information. The IT team wasn't explicitly searching for this, but they stumbled upon it one day when they were exporting a report from BetterCloud and noticed that the employee's posting frequency was four times higher than the next user. When they dug in and searched for specific words like *gun* in Arabic, they found even more terrorism-related posts. The employee was eventually arrested in the Philippines, which led to the arrest of several other ISIS-related individuals.

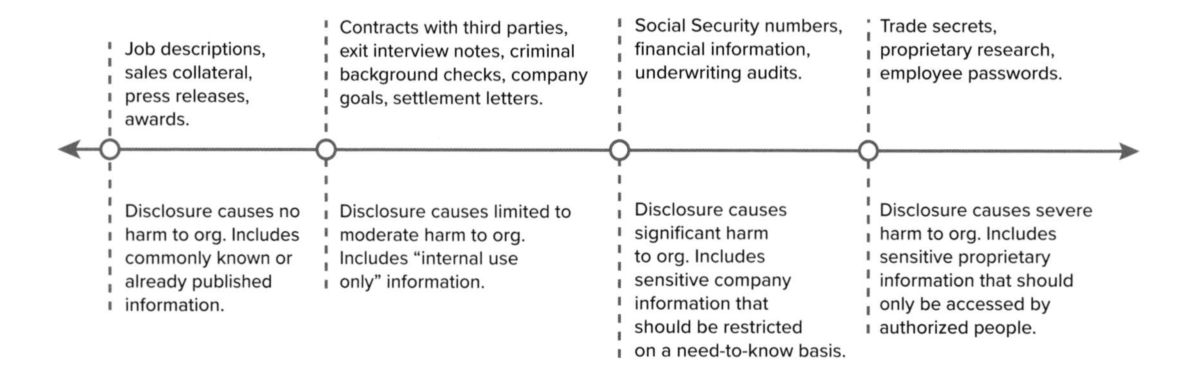

Figure 16: An example of a data classification scheme, illustrating how various types of data range from "not sensitive" to "sensitive."

Think about how you'd classify your organization's data. What's considered sensitive and what's not? If this data was disclosed, would it cause severe, significant, limited, or little to no harm to your organization?

It's the confluence of all these factors—identity, time, data sensitivity—that you need to think about when determining whether an interaction is trusted or not.

No single factor can define how risky an interaction is. It's the confluence of all these factors—identity, time, data sensitivity—that you need to think about when determining whether an interaction is trusted or not. Your definition of trust will vary based on the complexity of your IT environment, industry and compliance requirements, and business needs.

Given the sheer breadth of interactions, how do you secure them all? That's what we'll explore in the next chapter.

CHAPTER FIVE

The SaaSOps Philosophy Spectrum

CHAPTER FIVE HIGHLIGHTS

The SaaSOps Philosophy Spectrum

On one end of the **SaaSOps Philosophy Spectrum** are companies who believe in unfettered access and complete user trust. On the other end are companies who lock everything down in order to mitigate all risk.

Most modern workplaces fall somewhere in the middle. They want to secure their environments but also enable productivity.

Many existing IT and security tools only offer binary, **"all or nothing"** solutions. This creates a painful tradeoff between productivity and security.

That's why **it's critical to have flexibility and precision when securing interactions**. By having the freedom to be as lenient or strict as you want, thereby striking the right balance for your unique business needs, you no longer have to sacrifice security or employee productivity.

So what's the best way to secure all of these thousands, even millions, of interactions?

It depends who you ask. Everyone's approach varies.

Ask a four-person startup how they secure their digital workplace and you'll get a very different answer from what a 100-year-old enterprise company with 70,000 employees might give.

Here's what I call the SaaSOps Philosophy Spectrum. It illustrates the spectrum of philosophies:

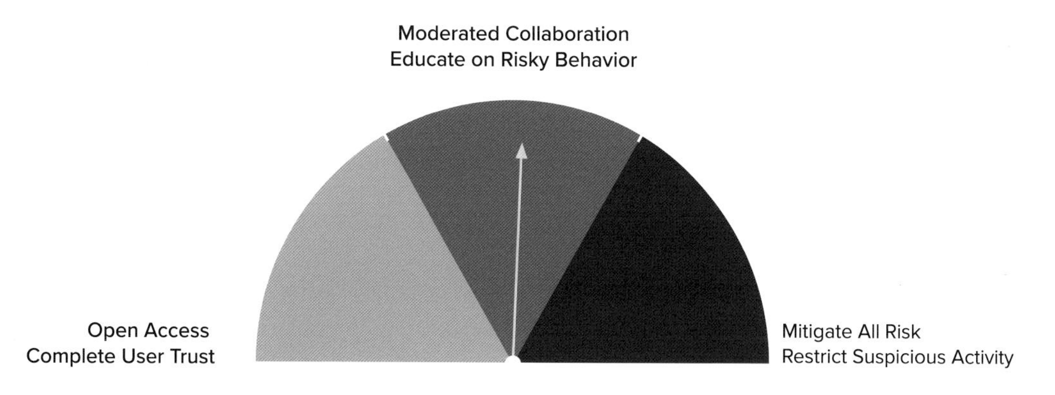

Figure 17: The SaaSOps Philosophy Spectrum illustrates the range of philosophies on how to secure user interactions.

Two opposing philosophies

On one end of the spectrum are companies who believe in unfettered access. Their IT teams have complete trust in their users. Every interaction is a trusted one, every policy an open one. Users can collaborate freely and easily with anyone they choose. They can invite people outside the org to view, comment on, or edit shared files. Nothing is restricted. There's little to no IT oversight.

On the other end of the spectrum are companies who take a much stricter approach. These are often organizations like banks and hospitals, which are heavily regulated and thus are required to enforce stringent security protocols.

To mitigate all risk, they lock everything down. To reduce data leaks and exfiltration risks, they might prohibit users from sharing data with anyone outside the org, whether through email attachments, invitations, links, etc. They might prevent users from installing third-party apps entirely. They might disable a variety of features, like the ability to create groups, create public links, access offline docs, publish files on the web, forward email automatically, etc. Any activity that could possibly be considered suspicious is restricted.

Neither approach is entirely realistic or ideal. The "open access" philosophy may not even be feasible if you're subject to compliance laws and regulations. Meanwhile, the "total security" approach is evocative of Gene Spafford's famous quote:

"The only truly secure system is one that is powered off, cast in a block of concrete, and sealed in a lead-lined room with armed guards—and even then I have my doubts."

While the "lock everything down" philosophy might seem tempting at first, it's too extreme for companies undergoing digital transformation.

While the "lock everything down" philosophy might seem tempting at first, it's too extreme for companies undergoing digital transformation. First, it often backfires. When users have a will, they'll find a way, even if that means circumventing IT to get their jobs done. By blocking everything, companies unintentionally drive their employees to adopt risky, unauthorized solutions. As Spafford's quote suggests, there's likely no such thing as total security.

Second, this philosophy defeats the very purpose of SaaS as it renders all the collaboration capabilities useless. Why migrate to the cloud and deploy SaaS applications if you can't reap the benefits? "I didn't deploy Box only to lock down sharing," one frustrated CISO told me.

Third, it often creates unwanted byproducts: new problems (and work) for IT. I recently spoke with the CIO of a global financial services company that had recently moved to SaaS. He told me that he adopted a "block everything" approach. Before long, his helpdesk was inundated with tickets from users requesting one-off access to data, permission to share data with others, elevated access, etc. It created hours of manual work for his team, leaving them with no easy way to track everything.

A painful tradeoff

It's a common conundrum in the digital workplace right now: How do you balance IT's needs for security and compliance with users' desires for frictionless collaboration?

If you let users have easy access to absolutely everything and give them unrestricted sharing capabilities, you'll end up risking a data breach or compliance violations. On the other hand, if you lock everything down, you'll end up hindering productivity (and likely encouraging shadow IT).

> How do you balance IT's needs for security and compliance with users' desires for frictionless collaboration?

With limited options offered by today's tools, you're forced to make a painful tradeoff.

Ultimately, you want to reconcile these two opposing pressures and find that equilibrium.

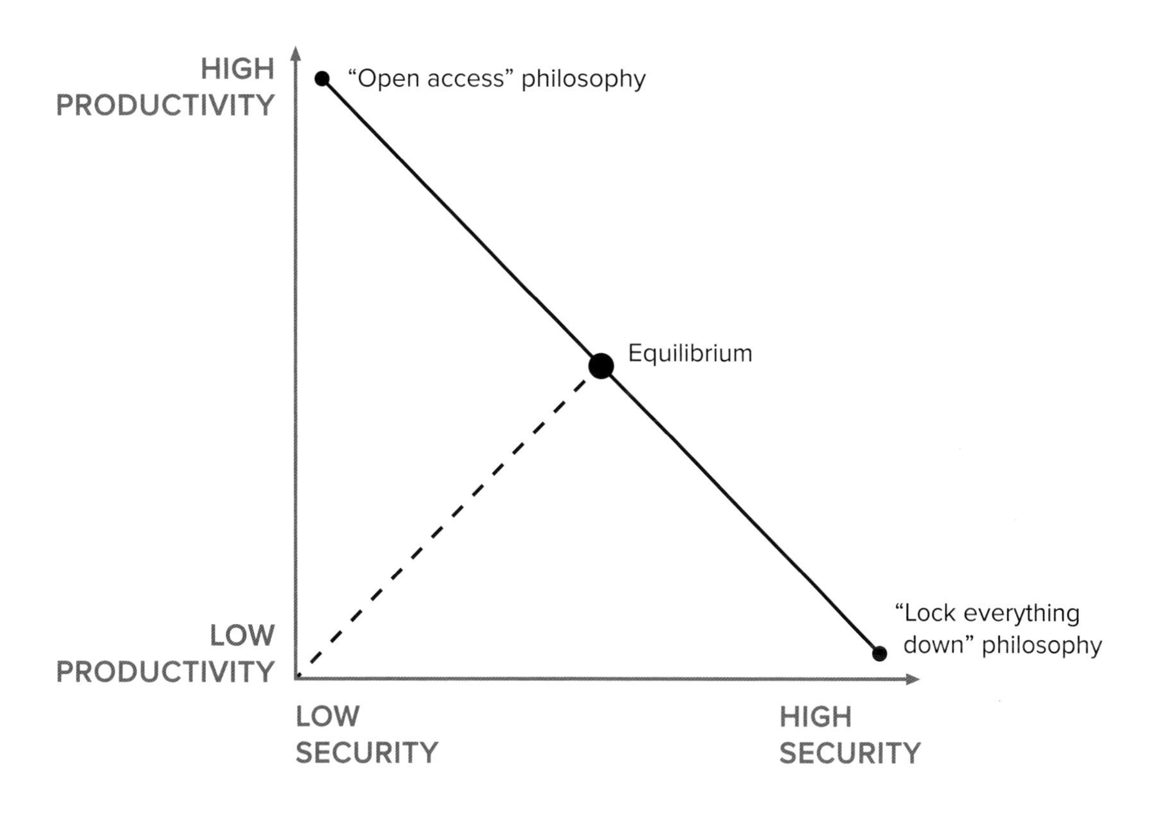

Figure 18: The tradeoff between productivity and security, and the equilibrium in the middle.

Finding the right balance

Most modern workplaces have a philosophy that falls somewhere in the middle of the productivity/security spectrum.

On a recent webinar poll, we found that only 3 percent of IT professionals adopted the "open access, complete user trust" philosophy, while 31 percent preferred the "mitigate all risk, restrict suspicious activity" route.[26] The majority (66 percent) took a more tempered approach: allowing moderated collaboration while educating users on risky behavior.[27]

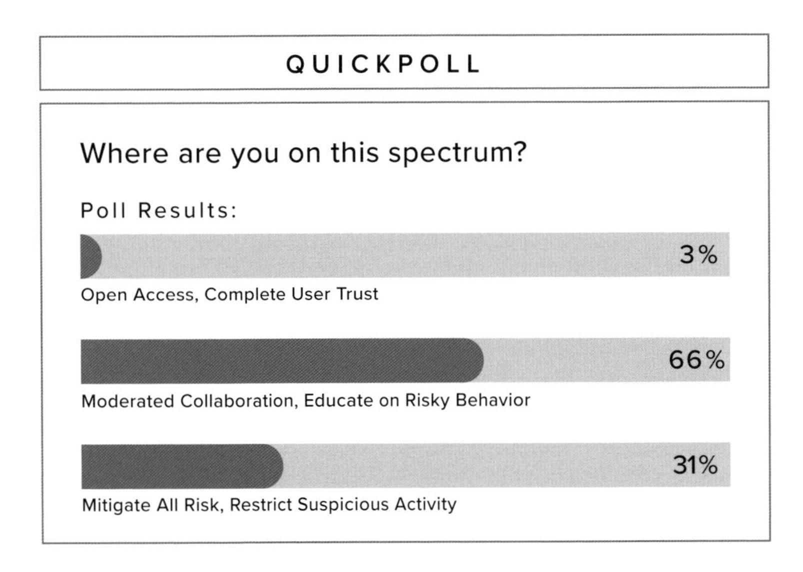

Figure 19: Results from a 2019 webinar poll showing where IT professionals land on the SaaSOps Philosophy Spectrum.

Having an approach that's somewhere in the middle of the road is the most effective way to ease the tension between the needs of the business and that of the user.

These IT teams want to enable collaboration and productivity, yet they also need the ability to mitigate risk and secure their SaaS environment. To do so, they prefer to educate their users on risky behaviors rather than lock everything down. When it comes to remediation, they prefer to take corrective action rather than simply block access.

For example, let's say a payroll specialist wants to share a sensitive payroll document with a few colleagues. She shares it as "Public on the web" to make it easier for collaborators to access. She means well, but doesn't realize the security implications of her actions. The

26 Buyers, Maddie. "Closing the Gap in SaaS Security." BetterCloud Monitor, 28 Mar. 2019, www.bettercloud.com/monitor/closing-the-gap-in-saas-security/.
27 Ibid.

IT team would prefer to automatically revert the sharing setting back to Private, email the user to explain why her behavior is risky, and send her manager a Slack message rather than completely restrict all file sharing for the entire company. Some IDaaS vendors even recommend shutting off a user's access entirely, though this remediation approach is likely to prove too extreme for most situations.

Or here's another example. Let's say you notice a pattern of suspicious logins from a legal assistant: She's repeatedly signing in from an unusual location. To be safe, you might want to temporarily suspend her account, resetting her sign-in cookies and OAuth tokens. But what if the same thing happened, except it was the CEO? Automatically applying the same exact policy might (unnecessarily) result in significant business disruption, not to mention a very peeved executive who was just trying to sign in while traveling abroad. In that case, you might want to automatically send a Slack message to an executive assistant to see what's going on before you start resetting the CEO's passwords.

These are just two example scenarios, but you get the idea: Solutions can't be one-size-fits-all. That's because trust is a fluid concept. Interactions have nuances. You need to be somewhere between the two extremes on the security/productivity spectrum. You need the flexibility to secure interactions differently based on all the dimensions we discussed earlier: identity, time, data sensitivity, and your own business requirements.

Solutions can't be one-size-fits-all. That's because trust is a fluid concept. Interactions have nuances.

But while many companies *want* to be in the middle of this spectrum, their desires are often thwarted. Many existing IT and security tools only offer sweeping solutions, such as blocking access or quarantining sensitive documents. This approach is often too broad and heavy-handed for digital workplaces. One IT professional we spoke to used this comparison: "With CASBs, we feel like we're trying to kill a small bug with a sledgehammer."

That's why it's critical to have flexibility and precision in how you choose to secure your organization's interactions. By having the freedom to be as lenient or strict as you want, thereby striking the right balance for your unique business needs, you no longer have to sacrifice security or employee productivity.

Note: It's important to remember that your philosophy may shift over time. What you want today may not be what you want in five years, one year, or even six months from now. Your company will change. It may scale (or shrink) in size and global presence, go public (or private), undergo an M&A (or divestiture), etc. The data privacy landscape will change. New data protection and compliance legislation will be enacted. You need to adopt a philosophy for the company you will become. Likewise, the tools you use should be flexible so that they can adapt as your company evolves.

What you want today may not be what you want in five years, one year, or even six months from now. You need to adopt a philosophy for the company you will become.

Where do you think your organization would fall on the SaaSOps Philosophy Spectrum? Next, I'll share several key factors to think about as you plot yourself on the spectrum.

CHAPTER SIX

Plotting Yourself on the SaaSOps Philosophy Spectrum

CHAPTER SIX HIGHLIGHTS

Plotting Yourself on the SaaSOps Philosophy Spectrum

Where you land on the SaaSOps Philosophy Spectrum depends on your **organization's risk tolerance, how your users use and share information** (both internally and externally), **and your company's requirements**.

Additional factors include **industry, size, company age, public status, M&A activity, infosec policies, turnover rate**, etc.

Take the quiz in this chapter to see where you fall on the SaaSOps Philosophy Spectrum.

Now that you've seen the SaaSOps Philosophy Spectrum, it's time to look at your own philosophy.

Where do you see yourself falling on this spectrum? What's your security stance?

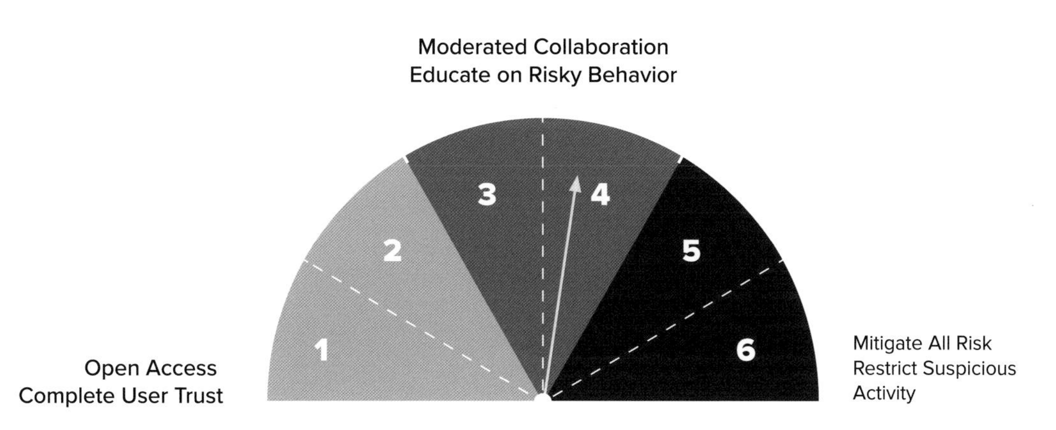

Figure 20: The SaaSOps Philosophy Spectrum, broken up into six zones. Where do you land?

Where you land depends on your organization's risk tolerance, how your employees use and share information (both internally and externally), and your company's requirements. In my experience working with 2,000+ digital workplaces, I've noticed a few key factors that can impact where you fall:

- **Industry.** Highly regulated industries like finance and healthcare tend to be on the right side of this spectrum since they're subject to strict regulatory oversight (examples: the Sarbanes-Oxley Act, GLBA, PCI DSS, HIPAA).
- **Size.** Smaller companies tend to fall on the left side of the spectrum, although that's not always true. It's easier to trust (and manage the data of) 10 employees than 10,000. As you grow, however, you may find your philosophy shifting to the right.
- **Company age.** Young startups fall more on the left side of the spectrum. Because many startups are born in the cloud, "open" policies feel like second nature. In contrast, for the 100-year old enterprise company that started on on-prem technology, SaaS apps are unfamiliar territory. Without the ability to duplicate the control they once had using on-prem systems, a common knee-jerk reaction is to lock everything down. It's not surprising to see them identify somewhere on the right side of the spectrum.
- **Internal vs. external collaborators.** Does your organization often collaborate with

external parties like partners or customers, or is work fairly contained? If your employees frequently share information with outside parties, this may place you on the right side of the spectrum.

- **Public status.** If you're a public company or you're preparing to IPO, this will also nudge you to the right on the spectrum. Public companies are subject to far more extensive reporting requirements, regulations, and public scrutiny, as well as regular audits. For example, all publicly traded companies in the US (and publicly traded non-US companies that do business in the US) must comply with SOX. Most businesses, regardless of size, setup, and business processes, will need security policies that include access controls, audit trails, data integrity controls, security assessments and audits, system authentication, system monitoring, and user provisioning to help with SOX compliance.[28]

How tightly the acquired companies are integrated will also affect how your SaaS instances are integrated, how tightly you secure them, and the flow of information between them.

- **M&A activity.** When you acquire a company, its cyber risk and IT environment become your responsibility to manage. Additionally, each industry and country has its own set of rules and laws. If you're diversifying into a new industry and/or the target company is located in a different country, this may change your security standards and policies. How tightly the acquired companies are integrated will also affect how your SaaS instances are integrated, how tightly you secure them, and the flow of information between them.

- **Infosec policies and procedures.** What are your existing corporate policies? For example, what are your policies for acceptable use, remote access, access control, information security, and email/communication? Do you allow personal devices to connect to your VPN? Do you allow or ban certain apps? Does your organization have a BYOD policy? All of this reflects your security maturity level and may influence where you fall on this scale.

28 "Keeping SOX 404 Under Control(s)." *SearchFinancialSecurity*, Jan. 2008, searchfinancialsecurity.techtarget.com/tip/Keeping-SOX-404-under-controls.

- **Previous data breaches or leaks.** A previous data breach, leak, or security incident may have led to tighter corporate security measures, stricter policies, more involvement from the C-suite, or revamped response and recovery plans, any of which could push you to the right on this scale. For example, in 2019, Google's top legal executive informed all Google employees that "accessing documents classified as 'need to know' without permission could result in termination . . . In the past, Google has had a reputation for openness, allowing employees wide access to documents and source code regardless of their job assignment. Now, following leaks about products in China and partnerships with the US military . . . Google is tightening the reins."[29]

- **Turnover rate.** The average turnover rate for all industries is 17.8 percent.[30] But if you're in an industry that has a higher-than-average turnover rate (e.g., hospitality or retail), that may shift you more to the right on the spectrum. You may prefer to enforce stricter rules given your employee churn rate.

- **Worker type (remote/part-time/temp vs. full-time).** Similarly, if you have a significant number of remote, part-time, or temporary employees (e.g., contractors), that may also place you on the right side of the spectrum. Less loyalty, higher turnover, a transient workforce, users relying on their own Wi-Fi to access company data, lack of control over user devices—all of this may mean stricter security rules.

- **Tech stack.** How heavily is your organization using technology that lends itself to these risks? Your tech stack may influence where you fall on the scale. For example, as companies adopt more SaaS applications, they feel increasingly vulnerable to insider threats. SaaS-heavy environments (defined as companies with 26 percent or more apps in the cloud) are 23 percent more likely than non-SaaS companies to feel vulnerable to insider threats.[31]

Less loyalty, higher turnover, a transient workforce, users relying on their own Wi-Fi to access company data, lack of control over user devices—all of this may mean stricter security rules.

29 O'Donovan, Caroline. "Google Exec's Internal Email On Its Data Leak Policy Has Rattled Employees." BuzzFeed News, BuzzFeed News, 14 May 2019, www.buzzfeednews.com/article/carolineodonovan/google-execs-internal-email-on-data-leak-policy-rattles.
30 "2016 Turnover Rates by Industry." Compensation Force, 21 Apr. 2017, www.compensationforce.com/2017/04/2016-turnover-rates-by-industry.html.
31 "State of Insider Threats in the Digital Workplace 2019." *BetterCloud Monitor*, 29 Apr. 2019, www.bettercloud.com/monitor/insider-threats-digital-workplace-2019/.

- **Location.** Your company may be subject to data privacy laws depending on where your company is based and where your customers are based. For example, sweeping regulations like GDPR impose stricter obligations for data security. In particular, they introduce new obligations for breach notification standards and expand the definition of "personal data" and "data breach." All of this has an operational impact for your IT team. You must know what data you have, where it lives, how it's protected, who has access to it, and what it's used for. With the California Consumer Privacy Act (CCPA) taking effect in 2020—and multiple states following suit with their own copycat bills—you may need to implement stricter data protection processes to secure customer data.

As you think through these factors and how they might influence your position on the spectrum, you may make some realizations. For example:

- You might be a "six" now, but you want to be a "four." Perhaps you're an enterprise company that's undergone a digital transformation recently. You've locked everything down, but your employees are demanding more flexibility and collaboration. You'd like to accommodate their way of working by making your policies less strict.

- You might be a "two" now, but in a year, you'll probably be a "five." Perhaps you're a fast-growing company that, up until now, has focused on scaling, not security. But there's an IPO on the horizon. As a result, you're anticipating tighter security controls, rigorous audits, and new reporting requirements, necessitating a move to the right side of the spectrum.

- You might think you're a "four," but in reality you're closer to a "two." Perhaps you realize your remediation approach isn't as strict as you thought, or perhaps you only have a few structured processes in place and you realize you need more.

A quiz to find out where you land on the scale

To help you determine where you fall along the SaaSOps Philosophy Spectrum, I've put together a quiz. My hope is that it sheds light on your current SaaSOps philosophy (or gets you thinking about where you want to be on the scale).

Note: There are probably hundreds of in-depth questions that could be asked, but I've purposely made these questions fairly simple. They're just meant to get you started and give you a general idea of where you fit on the spectrum.

Instructions: Give yourself **one** point for each question you answer "A," **two** points for every question you answer "B," and **three** points for every question you answer "C." Select the answer that most closely represents the action you'd take.

1. WHEN MANAGING SAAS APPLICATIONS, WHAT DO YOU CARE ABOUT *MOST*?

A) Enabling employees to be as productive as they can

B) Enabling productivity, but also securing sensitive or confidential data

C) Securing my company's sensitive or confidential data

2. IF YOU DISCOVERED THAT AN EMPLOYEE WAS REPEATEDLY FORWARDING CONFIDENTIAL CORPORATE EMAIL TO A PERSONAL GMAIL ADDRESS, WHAT WOULD YOU DO?

A) Ignore it; that doesn't matter to me

B) Email the user to explain why that's a security risk and inform their manager

C) Disable the automatic email forwarding capability for all users in the company

3. IF YOU DISCOVERED THAT A MANAGER EXPOSED A CONFIDENTIAL DOCUMENT PUBLICLY THROUGH A SAAS APP, WHAT WOULD YOU DO FIRST?

A) Email the user to inform him that he exposed a confidential document and trust him to remediate it on his own

B) Revert the sharing setting back to Private, remove collaborators, and email the user to explain why that's a security risk

C) Suspend the user's account and disable file sharing capabilities outside the org for all users in the company

4. IF A USER VIOLATES A POLICY ONCE, WHAT HAPPENS?

A) Nothing

B) They receive a warning and we explain to them why the policy violation is important

C) They're flagged as high-risk and their sharing access is locked down

5. WHAT DO YOU THINK OF GENE SPAFFORD'S FAMOUS QUOTE: "THE ONLY TRULY SECURE SYSTEM IS ONE THAT IS POWERED OFF, CAST IN A BLOCK OF CONCRETE AND SEALED IN A LEAD-LINED ROOM WITH ARMED GUARDS—AND EVEN THEN I HAVE MY DOUBTS"?

A) That approach seems excessive

B) Sounds about right. Despite your best efforts, you can never truly secure anything

C) You could add even more guards and wrap chains around it

6. HOW DO YOU FEEL ABOUT USERS INSTALLING THIRD-PARTY APPS (E.G., CHROME EXTENSIONS) THAT HELP THEM BE MORE PRODUCTIVE?

A) They can use third-party apps; we trust that they won't install anything malicious

B) They can use third-party apps, but we'd actively audit installations

C) No, third-party apps are too much of a security risk

7. DO YOU HAVE A FORMAL PROCUREMENT PROCESS FOR NEW SAAS APPS?

A) No, any departments can procure their own apps

B) No, but generally only IT can officially procure and deploy SaaS apps

C) Yes, users must go through a formal process involving IT, security, and finance teams

8. IF A CONTRACTOR WAS JOINING THE COMPANY ON A SIX-MONTH CONTRACT, WHAT SHOULD SHE BE PROVIDED WITH, IN YOUR OPINION?

A) A new email address (firstname.lastname-contractor@company.com), access to all public files in the company's cloud office suite, and a Member Slack account. This way she can find whatever she needs (and talk to whoever she needs) in order to get her job done

B) A Multi-Channel guest account in Slack and edit access only to the folders pertinent to her work

C) A Single-Channel guest account in Slack and read-only access to the files pertinent to her work (with the options to download, print, and copy disabled). Her access should be restricted to the bare minimum for security purposes

9. NOW THAT CONTRACTOR'S SIX-MONTH CONTRACT END DATE IS APPROACHING. WHAT ARE YOU PLANNING TO DO?

A) Nothing right away. Even if her contract ends, we trust that she wouldn't do anything dangerous

B) Email her manager and ask if access needs to be extended

C) Immediately suspend account once six months is up

10. IF YOU THOUGHT YOU HAD ONE SUPER ADMIN IN A SAAS APP BUT DISCOVERED THAT YOU ACTUALLY HAD 18, WHAT WOULD YOU DO?

A) Nothing. It probably helps people do their jobs faster because they don't need to continually request elevated access

B) Review who has admin access and find out why they need elevated privileges, revoke rights as necessary, and email them explaining why excess admin privileges are dangerous

C) Immediately revoke everyone's super admin access, keeping only one super admin to maintain the principle of least privilege

11. WHAT DO YOU THINK ABOUT EMPLOYEES USING THEIR OWN PERSONAL DEVICES FOR WORK PURPOSES?

A) We support the idea because it helps them be more productive; we don't need an MDM solution because we trust our users

B) We support the idea, but only if you employ an MDM for some device types

C) It's too risky unless you have a full MDM solution for *all* endpoint types across the company (and even then, it's questionable)

12. HOW DO YOU VIEW SHADOW IT?

A) It's not something I'm worried about

B) It's an opportunity to help employees be more successful; it identifies a gap between what IT currently supports and what end users need

C) It's something that should be immediately shut down once discovered; it's a security threat

13. AN END USER WANTS TO USE A TOOL THAT'S NEW ON THE MARKET. YOU'VE NEVER HEARD OF IT. HE STARTS EXPLAINING HOW IT'D BE A GAME-CHANGER FOR HIS JOB. WHAT'S YOUR GUT REACTION?

A) Yes, if it helps him work more efficiently

B) Maybe, let's do an assessment

C) No, this tool doesn't sound safe

14. IN YOUR OPINION, HOW DO IT TEAMS PROVIDE THE *MOST* VALUE?

A) Enabling the business to grow and be productive

B) Enabling growth and protecting corporate data

C) Ensuring that corporate data is protected

15. OF THE CHOICES BELOW, WHICH ONE IS THE *LEAST* IDEAL?

A) A user is restricted from sharing a file with a partner and has to wait five days for approval, resulting in lost productivity and a bad user experience

B) A user accidentally shares a sensitive file publicly within the domain, despite repeated training sessions and emails about the risks

C) A group containing confidential emails is exposed publicly, putting corporate data at risk

16. WOULD YOU EVER LET A USER SHARE THEIR CALENDAR WITH SOMEONE OUTSIDE THE ORG, IF IT HELPED THEM PLAN FOR AN UPCOMING PROJECT TOGETHER?

A) Yes, if it helps their productivity

B) Yes, but we'd want to monitor it to make sure it doesn't get publicly exposed

C) No, that's too risky; we want to keep all data inside the company

17. WHICH DEFINITION MATCHES YOUR DEFINITION OF A TRUSTED USER MOST CLOSELY?

A) Everyone in my org and anybody outside the org who needs to collaborate

B) Some people in my org and some people outside the org who need to collaborate

C) Some people in my org and very few people outside the org who need to collaborate

18. IN YOUR OPINION, WHAT'S THE BEST WAY TO MITIGATE INSIDER THREATS?

A) We trust our users; we think the risk of insider threats is low

B) Let them use SaaS apps with a few guardrails in place, but educate them on risky behavior

C) Severely restrict users' freedom; the less they can do/access, the smaller the chance of a data leak

19. WHAT'S YOUR PHILOSOPHY ON GROUP CREATION (E.G., GOOGLE GROUPS, SLACK CHANNELS, OFFICE 365 GROUPS)?

A) Users should be able to create their own groups as they see fit; it helps them be productive

B) Users can create groups, but IT should actively monitor them for any risky settings

C) Only IT should be able to create groups; this helps mitigate risk

20. IF YOU DISCOVERED THAT AN EMPLOYEE ACCESSED SENSITIVE INFORMATION THAT WASN'T RELEVANT TO THEIR JOB, WHAT WOULD THE APPROPRIATE COURSE OF ACTION BE?

A) Nothing; we trust that they won't do anything nefarious

B) Modify the sharing settings of the file; email the user to explain why this is bad behavior

C) Terminate the employee

21. WHICH STATEMENT BELOW ABOUT GROUPS IS THE MOST CONTROVERSIAL TO YOU?

A) Only IT can add users to groups, pending approval from the group owner, which may take up to a week

B) People can become group members only if they're invited by a group manager

C) We allow our users to view all group content and add themselves to groups without IT's approval

22. WOULD YOU EVER ALLOW PUBLIC GROUPS OR GROUPS THAT ALLOW EXTERNAL MEMBERS (E.G., WEB FORUMS, Q&A FORUMS, OR COLLABORATIVE INBOXES)?

A) Yes, without a doubt

B) Yes, but it depends on what they're being used for

C) No, we don't like the idea of any public groups, no matter what

23. IT TEAMS EXIST TO SUPPORT THE MISSION OF THE ORGANIZATION. HOW DO YOU DO THAT?

A) By reducing friction for users so that they can get their work done

B) By balancing the needs of the user with the needs of the business

C) By making sure corporate data is protected and by mitigating risk

24. A USER WANTS TO SHARE A SENSITIVE FILE OUTSIDE THE ORG WITH A FREELANCER. THEY ARE COLLABORATING ON A PROJECT TOGETHER. WHAT WOULD YOU DO?

A) Allow sharing without monitoring it; we trust that they'll behave appropriately

B) Allow sharing, but we'd monitor the file for any risky behavior (like data exposure)

C) Get more information: Is there really a strong business need to share this file with this person?

25. WHICH WORDS MOST CLOSELY DESCRIBE YOUR IT APPROACH?

A) Open, free

B) Flexible, adaptable

C) Cautious, conservative

Now tally up your points in the table below:

A RESPONSES (ONE POINT FOR EACH "A" RESPONSE)	B RESPONSES (TWO POINTS FOR EACH "B" RESPONSE)	C RESPONSES (THREE POINTS FOR EACH "C" RESPONSE)

If you had:

25–41 points: You're a one or two on the SaaSOps Philosophy Spectrum. You have few, if any, restrictions for your users. You favor an "open access" approach in order to enable productivity for your workforce. Your policies are fairly lax; in fact, you may not even have clear policies and standards defined in your organization.

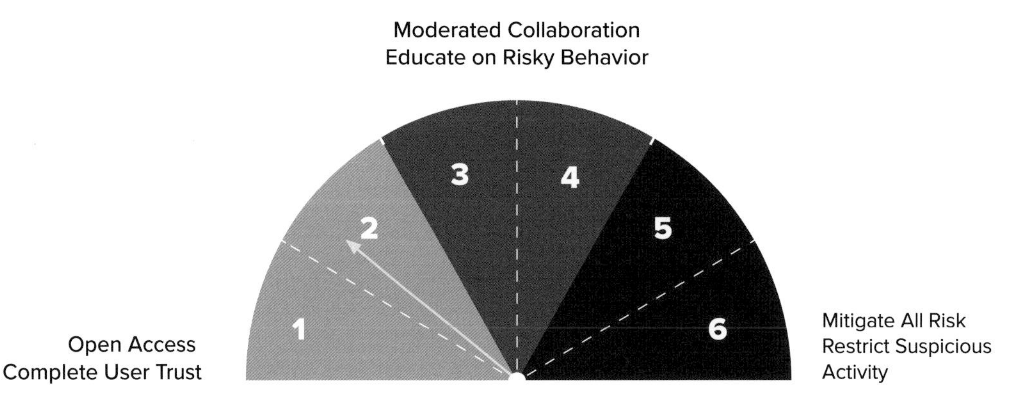

Figure 21: If you're a one or two on the scale, you favor an "open access" approach in order to enable productivity for your workforce.

42–58 points: You're a three or four on the SaaSOps Philosophy Spectrum. You take a more balanced approach in securing your SaaS apps. You want to enable productivity, but you also want to secure your data and mitigate risk. Some actions or behaviors may be restricted for users.

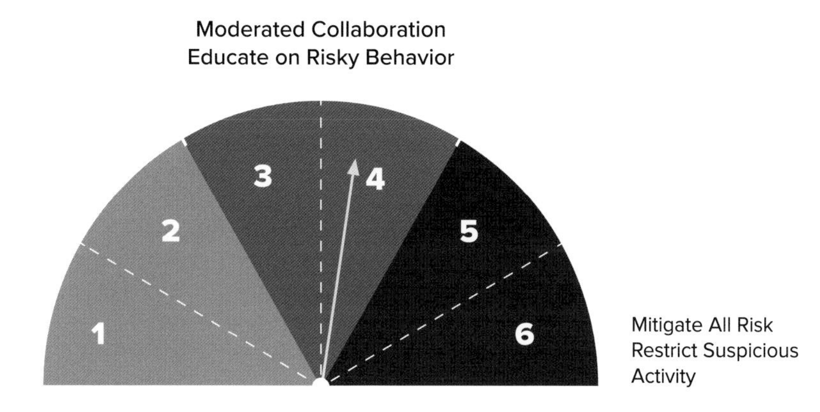

Figure 22: If you're a three or four on the scale, you take a more balanced approach in securing your SaaS apps.

59–75 points: You're a five or six on the SaaSOps Philosophy Spectrum. You value security above all else, even if it means lessening productivity for users. You have strict security policies and standards outlined in order to mitigate as much risk as possible. You may be in a highly regulated industry, like finance or healthcare.

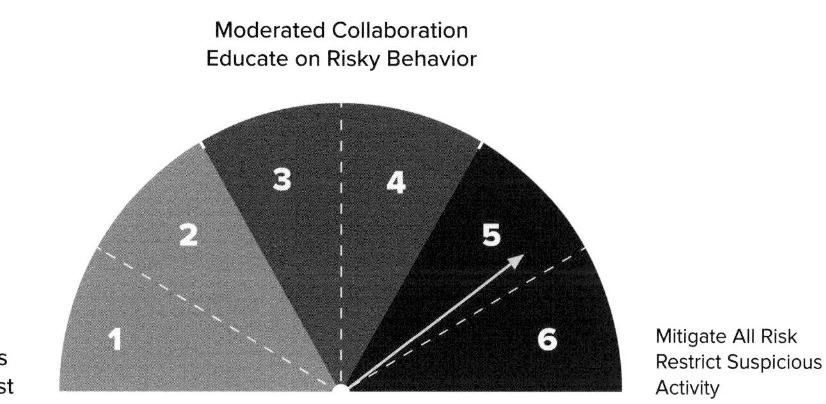

Figure 23: If you're a five or six on the scale, you have strict policies in order to mitigate as much risk as possible.

CHAPTER SEVEN

Solving Real-World SaaSOps Challenges

CHAPTER SEVEN HIGHLIGHTS

Solving Real-World SaaSOps Challenges

Here are several **true stories I've heard from companies who later became our customers.** These stories illustrate common, real-life IT challenges in the digital workplace. To preserve anonymity and protect privacy, names have been changed.

You'll learn how and why **data theft, excess admin permissions, improper offboarding, and misconfigured files** happen so easily when using SaaS apps.

You'll learn how each challenge can **impact a business and why it matters**.

You'll also learn **different approaches companies can take** based on their position on the SaaSOps Philosophy Spectrum.

In this chapter, I'll share some stories I've heard over the years from IT leaders who became our customers. To preserve anonymity and protect privacy, names have been changed. These stories are all true and illustrative of very real, very common IT challenges.

At the end of each story, I'll also outline how IT teams might respond to these situations based on their different positions on the SaaSOps Philosophy Spectrum.

1. Data theft

STORY #1: "IT'S MY BOOK OF BUSINESS"

Tim hung up the phone with the recruiter, grinning. That call had been the best one yet. His upcoming move was still a few months away, but he'd already started putting feelers out for a new job on the West Coast.

Casually, he logged into Salesforce. He had been a sales account executive at the company for two years and, in that time, he'd built up a hefty customer list. He thought of all the relationships he'd built, all the sales decks and proposals he'd put together. It'd be a shame to leave that all behind.

He eyed all of the reports he had set up in Salesforce. *Why shouldn't I download these?,* he thought to himself. *After all, it's my book of business. It's not like I'm stealing or anything. I nurtured and built these relationships. They're my customers. Why recreate the wheel at my next job?*

Hastily, he started downloading various reports: opportunities, leads, customer meetings, account details, contracts and orders, pipeline—anything he thought could potentially be useful at his next job. He was surprised how much data there was. He'd exported 50 different reports in the past five minutes.

He paused. Maybe, for good measure, he'd download a few sales decks and pricing lists from OneDrive too. These would come in handy if he ever wanted to show his new employer examples of his work.

Based on your stance on the SaaSOps Philosophy Spectrum, here are a few different ways you could respond to Tim's data theft:

Range of Remediation Approaches

PROFILE NUMBER ON THE SAASOPS PHILOSOPHY SPECTRUM	REMEDIATION ACTIONS
1 OPEN ACCESS, COMPLETE USER TRUST	IT does not track file sharing; trusts that users won't steal data
2	Email user to alert him of unusual download activity; trust that he'll stop on his own
3-4 MODERATED COLLABORATION, EDUCATE ON RISKY BEHAVIOR	Email user and user's manager to alert them of unusual download activity; open a ticket in ITSM with security team to investigate incident
5	Email user and user's manager to alert them of unusual download activity; send Slack message to security team; remove user as a collaborator from files; set up a policy that alerts IT if this happens again
6 MITIGATE ALL RISK, RESTRICT SUSPICIOUS ACTIVITY	Email user and user's manager to alert them of unusual download activity; send Slack message to security team; immediately suspend user in Salesforce and all file sharing platforms; block mobile devices until incident is resolved; disable IDaaS access across all apps

STORY #2: "THIS IS HOW YOU REWARD MY HARD WORK? WE'LL SEE ABOUT THAT"

Sighing, Natalie walked back to her desk, feeling crestfallen. Her boss had just delivered some bad news. The biotech company was preparing to lay off about 30 percent of the company, and she would be a part of this round of layoffs. Most of her department would be gutted. She had two months left.

Soon, she heard rumblings that a few colleagues were planning to jump ship and head to their biggest competitor, a massive pharmaceutical company. By now, her disappointment had turned to anger. Feeling emboldened, she applied to an open position. To her surprise, she heard back the same day. Calls turned into multiple rounds of interviews, which quickly materialized into a shiny job offer three weeks later.

Happily, she accepted. They settled on a start date just around the corner. But her resentment still lingered, even weeks later. The layoff news still stung. Hadn't she always gone above and beyond, devoted herself, and been willing to offer a helping hand? Hadn't she always received stellar performance reviews? And this is how her hard work would be rewarded? How could they do this to her?

We'll see about that, she thought. With steely determination, she opened up several folders in Dropbox. It was all there: her confidential clinical data and lab results. Biological summaries. Schematics. Data on new compounds being developed to treat cancer. Proprietary manufacturing protocols. She knew there were some breakthrough discoveries in here. And, more importantly, she knew leaking it to a competitor would deal a massive blow to her current company.

But she wouldn't download everything onto a USB stick. No, she'd do it a better way. One by one, she started sharing each file with her new work email address. This way, she'd always have access to the latest versions, even from her new job. Files, folders—she shared it all with herself. Her new boss was going to be floored by this goldmine.

Based on your stance on the SaaSOps Philosophy Spectrum, here are a few different ways you could handle this kind of data theft:

Range of Remediation Approaches

PROFILE NUMBER ON THE SAASOPS PHILOSOPHY SPECTRUM	REMEDIATION ACTIONS
1 OPEN ACCESS, COMPLETE USER TRUST	IT does not track file sharing; trusts that users won't steal data
2	Email user's manager to alert her of file sharing activity; trust that she'll remediate it on her own
3-4 MODERATED COLLABORATION, EDUCATE ON RISKY BEHAVIOR	Conduct one-time audit for any sensitive files shared with competitors' domains; if detected, remove collaborators and email user (or user's manager)
5	Regularly conduct audit for any sensitive files shared with competitors' domains; if detected, immediately remove collaborators, email user (or user's manager), send Slack message to security team to investigate incident
6 MITIGATE ALL RISK, RESTRICT SUSPICIOUS ACTIVITY	Regularly conduct audit for any files shared with competitors' domains; if detected, immediately remove collaborators, revoke user's access to applications, send Slack message to security team to investigate incident

WHY DATA THEFT MATTERS FOR YOUR BUSINESS

According to a report, 87 percent of employees said they take data they created with them when they leave a company.[32]

Many employees believe whatever they create during their tenure at a company is rightfully theirs. They feel ownership over their data and documents. According to the same report, 59 percent said they took data because they felt it was theirs, while 77 percent said they thought the information would be useful in their new job.[33]

Like Tim, these people may not think their behavior qualifies as theft. But if they've signed an employment contract or severance agreement stating that any documents created for the company are company property, there can be legal ramifications.

If they're taking confidential or sensitive corporate data with them, that's a significant problem, especially if they're going to a competitor. Recall the Zynga and Uber vs. Waymo lawsuits from Chapter Three. In both cases, an employee downloaded 14,000 extremely sensitive, highly confidential files and took them to a competitor. (Uber and Waymo eventually settled for $245 million after a year-long litigation process.)[34]

> If proprietary data ends up in the wrong hands, it can mean a potential loss of trade secrets, intellectual property, competitive advantage, market share, and revenue for your company.

The UK's communications watchdog, Ofcom, also suffered a serious data breach in 2016 when an ex-employee stole six years' worth of highly sensitive data and offered it to his new employer, an unnamed major broadcaster.[35] Or take the IT administrator at Columbia Sportswear who, after leaving the company, accessed his former employer's network 700 times, stole information, and turned it over to his new bosses. (He

32 Peters, Sara. "Survey: When Leaving Company, Most Insiders Take Data They Created." *Dark Reading*, 23 Dec. 2015, www.darkreading.com/vulnerabilities---threats/survey-when-leaving-company-most-insiders-take-data-they-created/d/d-id/1323677.
33 Ibid.
34 Marshall, Aarian. "Uber and Waymo Abruptly Settle For $245 Million." *Wired*, Conde Nast, 13 Feb. 2018, www.wired.com/story/uber-waymo-lawsuit-settlement/.
35 Fiveash, Kelly. "Ex-Ofcom Staffer Tried to Leak TV Company Data to Major Broadcaster." *Ars Technica*, 11 Mar. 2016, arstechnica.com/tech-policy/2016/03/ex-ofcom-staffer-tried-to-leak-tv-company-data-to-major-broadcaster/.

was later sued for malicious insider data theft.)[36]

These stories illustrate why IT teams need visibility into these kinds of red flag-raising interactions. If proprietary data ends up in the wrong hands, it can mean a potential loss of trade secrets, intellectual property, competitive advantage, market share, and revenue for your company.

2. Excess admin permissions

STORY: "IS THIS YOUR WAY OF TELLING ME I'M FIRED?"

Shannon was reaching her breaking point. She knew putting out fires was part of IT's job, but today felt particularly rough. One emergency after the other kept piling up.

She sighed. She didn't even know where to start.

Jay, her new helpdesk admin, looked at her. "Hey, did you see that ticket to recover those files? I think we have to restore the user to get that data back. He was only deleted a few days ago though, so we're good. The data's still there."

"I did see that ticket," she said. "Can you take care of it? They marked it as High Priority."

"Well, I think only super admins can restore deleted users, so you have to do it."

"Okay, one second. I'll deal with—"

They were interrupted by a knock on the door. Greg, the accountant, poked his head in.

"Hey guys, we have some auditors in the office. They're asking for domain admin access to perform their audits. Can you give that to them?"

"Well, I can export audit logs for them, but I'm pretty ba—"

36 Metzger, Max. "IT Admin Sued by Ex-Employer for Alleged Malicious Insider Data Theft." *SC Media*, 21 Mar. 2017, www.scmagazineuk.com/admin-sued-ex-employer-alleged-malicious-insider-data-theft/article/1475002.

Just then, the phone rang. “Hey, any luck with that file recovery? We really need it ASAP.”

Screw it, Shannon thought to herself. *It’s easier to just give them all access. I have a million things to do. They can handle things themselves and stop bothering me.*

In the Admin console, she hurriedly granted super admin access to everyone who needed it. She pulled up a spreadsheet and hastily jotted it down, making a mental note to take away Jay’s access the next morning and the auditors’ access in a month.

Unfortunately, she forgot all about it. She didn’t think about it again until six months later, when she found herself face to face with a wide-eyed, tense-looking Jay.

“Um, Shannon . . . I think I may have made a mistake,” he began nervously. “Don’t kill me. I was trying to do something in the Admin console and I think I might have, uh, deleted a bunch of users by accident . . . ”

“What? How?”

But before Shannon could properly answer, the phone rang—the first in a string of confused, angry phone calls.

“I can’t log in anymore.”

“What happened to my account?”

“Help! I have a really important client meeting in ten minutes and I can’t access anything at all!”

“Is this your way of telling me I'm fired?”

Shannon stared at Jay, panic rising within her. “Jay . . . what happened?!”

Based on your stance on the SaaSOps Philosophy Spectrum, here are a few different ways you could respond to Shannon and Jay’s admin fiasco:

Range of Remediation Approaches

PROFILE NUMBER ON THE SAASOPS PHILOSOPHY SPECTRUM	REMEDIATION ACTIONS
1 OPEN ACCESS, COMPLETE USER TRUST	IT does not track who has super admin access; trusts that any new super admins will act appropriately
2	IT is alerted for every new super admin added but takes no action
3-4 MODERATED COLLABORATION, EDUCATE ON RISKY BEHAVIOR	IT creates granular admin roles for anyone who needs elevated rights, including roles that automatically expire after a period of time (e.g., for auditors)
5	IT creates granular admin roles for anyone who needs elevated rights, including roles that automatically expire after a period of time (e.g., for auditors). They also set up a policy: There can only be three super admins in the org at all times. IT is alerted if an additional super admin is added; investigates if necessary
6 MITIGATE ALL RISK, RESTRICT SUSPICIOUS ACTIVITY	IT members all have two separate accounts: one normal user account and one privileged account. They also set up a policy: There can only be three super admins in the org at all times. If there are ever more than three super admins and this threshold is crossed, alert the IT and security team via Slack, automatically revoke the user's super admin access, and investigate the incident

WHY EXCESS ADMIN PERMISSIONS MATTER FOR YOUR BUSINESS

I've changed the names in this anecdote, but it's a true story. It comes from a company who later became one of our customers. Like many other IT teams, they had over-assigned elevated admin privileges. One day the helpdesk specialist accidentally deleted all the users in the sales org, resulting in a flurry of panic, anger, and confusion.

This isn't news to anyone: Least privilege is a security best practice. Don't dole out more privileges than necessary, and don't extend privileges longer than necessary. Only grant the bare minimum users need to do their jobs—nothing more, no less.

One day the helpdesk specialist accidentally deleted all the users in the sales org, resulting in a flurry of panic, anger, and confusion.

The least privilege model is effective because it reduces the number of ingress points.

According to *CSO Online*, "Every additional administrator causes linear-to-exponential growth in risk. Every additional admin doesn't just increase his or her own risk; if they're compromised, they add to the takedown risk of all the others. Each admin may belong to groups others do not. If a hacker compromises A and gets to B, B may more easily lead to C, and so on."[37]

The Edward Snowden case is an (extreme) illustration of what can happen when someone has excess admin permissions. An NBC investigation found that as a system administrator, Snowden was allowed to look at any file he wanted, and his actions were largely unaudited.[38] His contractor job "gave him unrestricted access to a mountain of sensitive materials for which he had no legitimate need."[39] Lesson learned: In 2013, the NSA eliminated 90 percent of its system administrators to reduce the number of people with access to secret information.[40]

37 Grimes, Roger A., and Roger A. Grimes. "Too Many Admins Spoil Your Security." *CSO Online*, CSO, 7 May 2013, www.csoonline.com/article/2614271/too-many-admins-spoil-your-security.html.
38 Esposito, Richard, and Matthew Cole. "How Snowden Did It." *NBCNews.com*, NBCUniversal News Group, 26 Aug. 2013, investigations.nbcnews.com/_news/2013/08/26/20197183-how-snowden-did-it.
39 Seltzer, Larry. "How Snowden Got the NSA Documents." *ZDNet*, 26 Aug. 2013, www.zdnet.com/article/how-snowden-got-the-nsa-documents/.
40 Allen, Jonathan. "NSA to Cut System Administrators by 90 Percent to Limit Data Access." *Reuters*, Thomson Reuters, 9 Aug. 2013, www.reuters.com/article/us-usa-security-nsa-leaks/nsa-to-cut-system-administrators-by-90-percent-to-limit-data-access-idUSBRE97801020130809.

Unfortunately, in SaaS environments, enforcing the least privilege model is easier said than done. Most SaaS apps, particularly the newer ones, don't offer admin roles with much granularity. There are very few options between super admin and end user.

As a result, orgs often end up with a glut of super admins across SaaS apps. IT may not necessarily want to give employees carte blanche, but they often have no choice. They have to either hand over all the keys to the kingdom or thwart employee productivity.

> IT may not necessarily want to give employees carte blanche, but they often have no choice.

Additionally, there's no easy way to keep track of, much less automate, admin access management over time in native admin consoles. This results in privilege creep. As employees change roles and gain access to new systems, they retain privileges to old ones. It happens to contractors and auditors as well. They often retain their elevated privileges long past their contract end date because their access must be revoked manually.

On a recent webinar poll, 68 percent of IT professionals said they're manually managing admin privileges (e.g., using spreadsheets, reviewing users one by one, etc.) across SaaS applications, while 5 percent said they aren't managing admin access at all.[41] When admin permissions aren't managed effectively, employees can retain alarming amounts of power for weeks, months, perhaps even years at a time, resulting in an increased attack surface.

41 Buyers, Maddie. "4 Insider Threats That Should Keep You Up at Night." *BetterCloud Monitor*, 13 June 2019, www.bettercloud.com/monitor/4-insider-threats-webinar-recap/.

3. Incomplete offboarding

STORY: "HE'D BEEN QUIETLY USING THIS INTEL TO UNDERCUT HIS FORMER FIRM"

Alex couldn't believe his luck.

It had been two years since he'd left his previous company, a VC firm, to become a partner at a rival firm. There had been one small but very important oversight—and *still* nobody had caught on. He'd thought that by now, surely, someone would have realized their mistake.

But no, they hadn't. Much to his delight, he could still log into his former firm's Dropbox account. His username and password still worked. There, in Dropbox, he could view everything the firm was working on: term sheets, investment plans, business valuation studies, due diligence documents, stock purchase agreements, voting agreements . . .

The IT team had never bothered to shut down his account, and their gaffe was his gain.

To say that this had served him well was an understatement. Over the course of two years, he'd been quietly using this intel to undercut his former firm.

Every time Alex was competing for a deal, he'd log into his old Dropbox account and find the term sheet. He'd scrutinize the terms and conditions his former firm was offering. Then he'd put together a new sheet with better terms and swiftly approach the relevant entrepreneur with it. Without fail, these tactics helped him win deal after deal, leading to a steady string of successful investments.

And now, he was looking at what might be the biggest, most important investment of the year. He *had* to win this deal; it'd be a home run for his firm. Smiling to himself, he entered his old login credentials and began scrolling through the documents.

Based on your stance on the SaaSOps Philosophy Spectrum, here are a few different ways you could handle offboarding, in order to prevent this kind of incident from happening:

Range of Remediation Approaches

PROFILE NUMBER ON THE SAASOPS PHILOSOPHY SPECTRUM	REMEDIATION ACTIONS
1 OPEN ACCESS, COMPLETE USER TRUST	IT offboards user when they get around to it; trusts that ex-employees won't abuse their access
2	IT resets password to lock user out of account but takes no further action; trusts that user won't try to access data another way
3-4 MODERATED COLLABORATION, EDUCATE ON RISKY BEHAVIOR	IT automatically transfers all files to new owner; automatically deprovisions user and revokes access across all SaaS apps using a checklist
5	Departing employee notification in HRIS automatically triggers offboarding workflow, which includes the following actions: reset user's password and sign-in cookies, remove user from groups, delete 2-step verification backup codes, transfer all files and primary calendar events, deactivate necessary licenses, and suspend user
6 MITIGATE ALL RISK, RESTRICT SUSPICIOUS ACTIVITY	Departing employee notification in HRIS automatically triggers offboarding workflow, which includes the following actions: reset user's password and sign-in cookies, disable IMAP/POP settings, turn off email forwarding, remove user from groups, delete 2-step verification backup codes, hide user in directory, remove user from shared calendars, revoke all apps, remove devices and wipe account, transfer all files and primary calendar events, deactivate necessary licenses, send Slack integration logs to security team, set auto-reply directing emails to user's manager, wait 30 days, and suspend user

WHY OFFBOARDING MATTERS FOR YOUR BUSINESS

Ex-employees who aren't offboarded correctly can retain access and take sensitive data to a competitor, as seen in the story above. Or, worse yet, they can wreak havoc on the business.

"First of all, we want to offer our apologies for any inconvenience. Unfortunately, an ex-administrator has deleted all customer data and wiped most servers. Because of this, we took the necessary steps to temporarily take our network offline."

That was a message posted on verelox.com, a virtual and dedicated server provider, after an improper offboarding incident. The site was down for over 30 days.

Many companies fail to offboard employees immediately—or, if they do, it's often incomplete. A 2017 survey found that 25 percent of IT decision makers take more than a week to deprovision a former employee, while 25 percent didn't know how long accounts remained active once an employee had left the company.[42] They also found that failure to deprovision employees from corporate applications caused a data breach at 20 percent of the companies represented in the survey.[43] Another report found that 39 percent of large businesses take up to a month to close dormant accounts.[44]

All of this data echoes what we found too. On a recent webinar poll, we learned that 76 percent of IT professionals believe that former employees still have access to their organization's data.[45]

As a result, horror stories about improper offboarding abound.

Take Marcelo Cuellar for example. Fired from a logistics and shipment company in 2015, he left to work for a competitor. In 2016—more than a *year* after his termination—he accessed his previous company's Google Drive account from his home. He proceeded to download over 1,900 spreadsheets created by employees at his previous company. Once the company discovered his illicit activity, they sued him for violating the federal Computer

42 "New Research from OneLogin Finds over 50% of Ex-Employees Still Have Access to Corporate Applications." *OneLogin*, 13 July 2017, www.onelogin.com/press-center/press-releases/new-research-from-onelogin-finds-over-50-of-ex-employees-still-have-access-to-corporate-applications.
43 Ibid.
44 "59% Of UK Workforce on Lookout for Jobs in 2016, Businesses Urged to Put Strict Access Controls in Place." *Ilex International*, 23 Feb. 2016, www.ilex-international.com/en/59-of-uk-workforce-on-lookout-for-jobs-in-2016-businesses-urged-to-put-strict-access-controls-in-place/.
45 Wang, Christina. "The Top Security Blind Spots in Your SaaS Environment." *BetterCloud Monitor*, 14 May 2018, www.bettercloud.com/monitor/the-top-security-blind-spots-recap/.

Fraud and Abuse Act (CFAA).[46]

The stakes are even higher if the ex-employees are IT admins. There are stories of disgruntled admins wiping out payroll files,[47] annihilating AWS servers,[48] slipping porn into their ex-CEO's PowerPoint presentation,[49] deleting patient health information and medical records,[50] completely shutting down production at a Georgia-Pacific paper mill,[51] and crashing an ISP network after leaving their companies.[52]

In these scenarios, the companies lost clients, lost major contracts, and experienced serious damage to their reputations. In the case where patient health information was deleted, the sysadmin had retained privileged access for *two years* after he was fired; his actions caused at least $5,000 in damages and potentially interfered with the diagnosis and treatment of patients.[53] In the case where AWS servers were deleted, police said the wreckage caused an estimated loss of $700,000. The company reportedly was never able to recover the deleted data.[54] In Georgia-Pacific's case, the ex-employee was forced to pay $1.1 million in damages from the downtime his actions caused.[55]

But fully offboarding departing employees in the digital workplace has become quite complicated.

The sheer volume of SaaS applications creates offboarding difficulties. SaaS data lives in multiple siloed applications, some sanctioned by IT and some not. Data sprawl across applications creates complexity. The rise of freelancers, contractors, and consultants adds another layer of complexity, since their access to data and applications is only temporary.

46 Shinn, Jason. "Ex-Worker Sued for Accessing Former Employer's Google Drive Account." *Michigan Employment Law Advisor*, 3 Apr. 2017, www.michiganemploymentlawadvisor.com/technology-employment-issues/computer-fraud-and-abuse-act/ex-worker-sued-accessing-former-employers-google-drive-account/.
47 McMillan, Robert. "Ex-Employee Hacks US Military Contractor's Computer Systems." *Techworld*, 5 Sept. 2011, www.techworld.com/news/security/ex-employee-hacks-us-military-contractors-computer-systems-3301315/.
48 "Sacked IT Guy Annihilates 23 of His Ex-Employer's AWS Servers." *Naked Security*, 23 Mar. 2019, nakedsecurity.sophos.com/2019/03/22/sacked-it-guy-annihilates-23-of-his-ex-employers-aws-servers/.
49 Matyszczyk, Chris. "Disgruntled IT Guy Slips Porn into CEO's PowerPoint." *CNET*, CNET, 22 June 2011, www.cnet.com/news/disgruntled-it-guy-slips-porn-into-ceos-powerpoint/.
50 Blake, Andrew. "Health Care Facility Hacked by Ex-Employee Using 2-Year-Old Credentials: Justice Department." *The Washington Times*, The Washington Times, 22 Mar. 2017, www.washingtontimes.com/news/2017/mar/22/healthcare-facility-hacked-ex-employee-using-2-yea/.
51 Cimpanu, Catalin. "Revenge Hacks Cost Former Employee 34 Months in Prison, $1.1 Million in Damages." *BleepingComputer*, BleepingComputer.com, 17 Feb. 2017, www.bleepingcomputer.com/news/security/revenge-hacks-cost-former-employee-34-months-in-prison-1-1-million-in-damages/.
52 Cimpanu, Catalin. "Sysadmin Gets Two Years in Prison for Sabotaging ISP." *BleepingComputer*, BleepingComputer.com, 4 Dec. 2016, www.bleepingcomputer.com/news/security/sysadmin-gets-two-years-in-prison-for-sabotaging-isp/.
53 Blake, Andrew. "Health Care Facility Hacked by Ex-Employee Using 2-Year-Old Credentials: Justice Department." *The Washington Times*, The Washington Times, 22 Mar. 2017, www.washingtontimes.com/news/2017/mar/22/healthcare-facility-hacked-ex-employee-using-2-yea/.
54 "Sacked IT Guy Annihilates 23 of His Ex-Employer's AWS Servers." *Naked Security*, 23 Mar. 2019, nakedsecurity.sophos.com/2019/03/22/sacked-it-guy-annihilates-23-of-his-ex-employers-aws-servers/.
55 Cimpanu, Catalin. "Revenge Hacks Cost Former Employee 34 Months in Prison, $1.1 Million in Damages." *BleepingComputer*, BleepingComputer.com, 17 Feb. 2017, www.bleepingcomputer.com/news/security/revenge-hacks-cost-former-employee-34-months-in-prison-1-1-million-in-damages/.

The offboarding process is also highly manual and time-consuming, leaving ample room for human error and oversight.

Offboarding affects the bottom line, whether it's through the risk of data breaches, failed compliance, hindered productivity, data loss, or loss of revenue. But the vast majority of companies only think about offboarding when it becomes a problem, not before. Knowing the risks and taking preventative and proactive actions will go a long way toward protecting your data.

4) Data exposure

STORY #1: "THAT SEEMS MUCH MORE CONVENIENT. IT'LL MAKE EVERYONE'S LIVES EASIER"

Rob perused the folder in Box he had just put together. It was comprehensive: It contained nearly 2,000 Word documents. Every document had each patient's name, home address, driver's license number, email address, Social Security number, and stay history recorded.

He worked at one of the biggest addiction rehabilitation centers in the country. His boss, the center's addiction specialist, had asked him to compile all of the patient data into one centralized folder. It had taken him weeks, but he was finally done.

Soon, word got out that he'd put together all of the patient files. The next day, he started receiving emails left and right requesting access.

"I need all the files," said the director of nursing.

"Could you send me the files for all new patients from July 1?" asked the billing specialist.

"It'd be helpful for me to have download access for the patients in the residential program," said the patient care coordinator.

"I sent the link to my assistant but it didn't work. She said she needs permission," said the director of therapy. "Can you share it with her?"

Rob, flustered, was already late to his next meeting. *I really don't have time to keep sharing this folder with people one by one,* he thought. *This is more trouble than it's worth.*

He resignedly started typing in his colleagues' names to share the folder with them.

But then he noticed a "Share Link" setting. It read underneath, "Publicly accessible and no sign-in required." *Well, that seems much more convenient,* he thought. *It'll make everyone's lives easier.*

Brightening, he enabled that option and headed to his meeting—completely unaware that 2,000 patients' PII and stay histories at the addiction rehab center were now publicly accessible.

Based on your stance on the SaaSOps Philosophy Spectrum, here are a few different ways you could respond to Rob's file sharing snafu:

Range of Remediation Approaches

PROFILE NUMBER ON THE SAASOPS PHILOSOPHY SPECTRUM	REMEDIATION ACTIONS
1 OPEN ACCESS, COMPLETE USER TRUST	IT does not track file sharing; trusts that users are sharing data appropriately
2	Email user to tell them that their file is publicly shared (no action required); trust that they will remediate it on their own
3-4 MODERATED COLLABORATION, EDUCATE ON RISKY BEHAVIOR	Scan all publicly shared files for sensitive information. If there is a publicly shared file containing sensitive information, revert it back to Private and email user and user's manager to inform them, explaining why it's a security risk
5	Scan all publicly shared files for sensitive information. If there is a publicly shared file containing sensitive information, revert it back to Private; email user and user's manager to inform them; change file owner to security manager; disable download, print, and copy permissions on file; send #security-alerts Slack channel a message; set up a policy that remediates this scenario if it happens again
6 MITIGATE ALL RISK, RESTRICT SUSPICIOUS ACTIVITY	Disable ability to share any files publicly/outside of domain in Slack, G Suite, Box, Dropbox, etc.

STORY #2: THE CHINESE WALL

Sydney looked up from the email she was writing. Her co-worker, Aaron, was standing by her desk, and he did not look pleased. She wondered briefly if their company, a financial services firm, was downsizing their IT department, and steeled herself.

"I just got an email from the compliance team. They need to talk to us later today," he said grimly.

"About what?"

"The wealth management team is getting a whiff of what the mergers team is working on. They should *not* know that. You know, 'Chinese Wall' and all that. But somehow they do."

"Oh," said Sydney, slightly relieved. "But they sit separately, on different floors. They can't see each other's screens. They can't overhear information."

"Well, some kind of wall-crossing is happening, and it can't be. The whole point of the Chinese Wall is to prevent conflicts of interest and insider leaks. Don't they use Box?"

"They do. Everyone's files are stored in one instance."

"And they can't see each other's files, right?" asked Aaron.

"Well, no. Not if the permissions are set up correctly. Only the people on each team can see their team's files," she said. "But I guess, technically, they could change those settings on their own . . . or if someone switches teams and we're not aware of it, I could see how this might happen . . . "

"We could get majorly fined for this," hissed Aaron. "We have to figure something out."

Based on your stance on the SaaSOps Philosophy Spectrum, here are a few different ways you could handle the Chinese Wall situation:

Range of Remediation Approaches

PROFILE NUMBER ON THE SAASOPS PHILOSOPHY SPECTRUM	REMEDIATION ACTIONS
1 OPEN ACCESS, COMPLETE USER TRUST	IT does not track file sharing; trusts that users are sharing files appropriately within company
2	Scan all files shared publicly within the domain for sensitive information. If any are shared, email user to explain why this is a security risk and trust that they will remediate it themselves
3-4 MODERATED COLLABORATION, EDUCATE ON RISKY BEHAVIOR	Scan all files shared publicly within the domain for sensitive information. If any are shared, email user to explain why this is a security risk and modify privacy settings
5	If a sensitive file is viewed/downloaded by a user in a specific department (who should not have access), then immediately email user, email user's supervisor, and message security and compliance team to investigate
6 MITIGATE ALL RISK, RESTRICT SUSPICIOUS ACTIVITY	If a sensitive file is viewed/downloaded by a user in a specific department (who should not have access), then immediately suspend user across apps, block mobile devices, email user's supervisor, and message security and compliance team to investigate

WHY MISCONFIGURED/EXPOSED FILES MATTER TO YOUR BUSINESS

Misconfigured files mean your corporate data is exposed in some way, shape, or form.

Both of the stories I just told are true. Both illustrate how data exposure (either within or outside of your domain) can result in serious security incidents. The first incident happened to a company that later became one of our customers. They were shocked and horrified to discover that the health information of thousands of their patients was shared publicly. But this is not an uncommon incident—SaaS makes it very easy for a well-meaning but negligent user to misconfigure settings, unaware of the security implications.

The second incident happened to a financial services company that had recently deployed Box. When they discovered that the research and underwriting arms could access each other's data within Box, they came to us for help building a better "Chinese Wall." They were struggling to find a way to secure Box files while also enabling productivity.

If your data is somehow exposed and untrusted users are accessing it, there can be serious consequences. This is especially true if you're in a highly regulated industry like finance. For example, in 2016, stockbroker and wealth manager WH Ireland was fined £1.2 million by the UK Financial Conduct Authority for its "weak catalog of controls," including an "ineffective information-sharing barrier, known as a 'Chinese Wall,' between the public-facing and private sides of its business."[56] Its corporate broking division was also banned from taking on clients for 72 days. This brings me back to my point about trusted vs. untrusted users in Chapter Three. Just because someone belongs to your org, it doesn't automatically qualify them as a trusted user.

These stories illustrate the different types of files that can be publicly shared. For example, files may be public in your domain, meaning anyone in your organization can find and access them, or they may be "Public with a link," meaning anyone at all with the link can access the file. Yet another option is "Public on the web," meaning the files are indexed by Google and can be found and accessed by anyone on the internet. No

56 Brown, John Murray. "WH Ireland Fined £1.2m by the FCA for Market Abuse Failures." *Financial Times*, Financial Times, 23 Feb. 2016, www.ft.com/content/3a95d25c-da03-11e5-a72f-1e7744c66818.

sign-in is required. This is the riskiest sharing setting. Anyone on the internet could potentially find, view, and download your confidential documents, spreadsheets, presentations, photos, or calendars.

That's what happened in 2019. Security researchers discovered major tech companies were leaking sensitive corporate and customer data because employees were sharing public links to files in their Box enterprise accounts. They found tax documents, customer lists, passwords, patient names, insurance information, non-disclosure agreements, and employee performance metrics.[57] Even worse, some public folders had been scraped and indexed by search engines, making it easier to find the data.

Publicly shared files are extremely risky since anyone on the internet (or domain) can find and access them. But if you're a "one" on the SaaSOps Philosophy Spectrum, maybe your open policy means that you don't care about that. You're okay with allowing wide access to data, regardless of job title or business unit. You trust your users. You're promoting a culture of openness. But if you're a "five" or "six," you likely don't want ***any*** employees to have any semblance of access unless it's critical for their job.

Are your employees using sharing settings for files that they shouldn't be? Having visibility into these interactions is critical for data protection. It protects critical IP, helping prevent loss of trade secrets, data leaks, negative PR, loss of consumer trust, compliance violations and fines, and more.

These are just a few of the biggest challenges in the digital workplace. There are more: 2FA enforcement, publicly exposed groups, users forwarding corporate email to personal accounts . . . but you get the gist of these difficulties. At the root of it, they all stem from user interactions.

Managing and securing user interactions might seem like a daunting task but in the next chapter, I'll outline an action plan that'll help you get started.

57 Whittaker, Zack. "Dozens of Companies Leaked Sensitive Data Thanks to Misconfigured Box Accounts." *TechCrunch*, TechCrunch, 11 Mar. 2019, techcrunch.com/2019/03/11/data-leak-box-accounts/.

CHAPTER EIGHT

Get Started with a SaaSOps Checklist & Action Plan

CHAPTER EIGHT HIGHLIGHTS

Getting Started with a SaaSOps Checklist & Action Plan

Managing and securing thousands of interactions day to day across the digital workplace might seem like a Herculean task.

This chapter contains a **SaaSOps checklist that you can start tackling today**, as well as a **long-term action plan for securing your SaaS data**.

With this plan, you can **start establishing IT policies, guidelines, processes, and standards** that reflect your definition of trust and align with company-wide organizational priorities.

Managing and securing tens of millions of interactions day to day across the digital workplace might seem like a Herculean task.

To start with, there are key operational processes in your SaaS environment to review, consider, and implement. Below is a best practices checklist for SaaSOps that you can begin tackling today:

- ☐ Check which external users are in your groups/distribution lists. Are there any users who no longer belong (e.g., contractors whose contracts have ended)?
- ☐ Check if there are any external domains or people your files are shared with
- ☐ Check who's forwarding email outside your domain
- ☐ Check your group privacy settings. Are any groups overexposed? ("Groups" can mean email lists, web forums, Q&A forums, collaborative inboxes, etc.)
- ☐ Check your calendar privacy settings. Are any calendars overexposed?
- ☐ Proactively monitor your files for risky events, such as:
    - ☐ Sensitive files shared publicly or externally
    - ☐ Sensitive folder paths (e.g., accounting or finance) shared publicly or externally
    - ☐ Sensitive files shared with a competitor or personal email account
    - ☐ Sensitive data exposed by executives (e.g., CEO, CFO)
    - ☐ Specific file types shared publicly or externally. Based on what I've seen with our customers, spreadsheets and PDFs usually contain the most sensitive information
    - ☐ Unusually large file downloads in a short window
- ☐ Scan your files regularly for PII, PHI, payment information, passwords, executable files, keywords that might signal sensitive information (e.g., "Confidential," "Internal Use Only"), and/or any confidential project names to make sure none of this is publicly exposed
- ☐ Make sure you have a robust and reliable offboarding process so that ex-employees don't have lingering access to corporate data

- ☐ Make sure you have a defined offboarding process for temporary workers, contractors, and vendors so that they don't have lingering access to corporate data after their time is up

- ☐ Consider creating step-by-step wizards for helpdesk admins to guide them through onboarding, offboarding, and any other multi-step process, so that no step is overlooked

- ☐ Consider who you want to transfer data to during the offboarding process and where that data lives

- ☐ Consider how you'll audit the offboarding procedure and where those logs live

- ☐ Review any user-based dependencies. In other words, which systems are tied to specific accounts and will break if that account is suspended or that user leaves the company?

- ☐ Check how many super admins you have across SaaS apps and reduce that number if necessary

- ☐ Make sure you're notified of any spikes in suspicious or failed user logins so that you can proactively reach out to the user(s) in question to investigate

- ☐ Enforce the principle of least privilege across SaaS apps. Create granular access roles so that users only have the bare minimum privileges necessary to do their jobs

- ☐ Consider limiting roles by time (e.g., a helpdesk admin only has rights Monday – Friday, from nine a.m. to five p.m.). Doing so will help enforce the principle of least privilege by preventing after-hours or weekend access

- ☐ Ensure all users have MFA enabled. Make sure you have a way of being notified if a user disables it

- ☐ Audit users who have active SaaS app licenses but haven't logged in in over 90 days. This can indicate users who should have been (but never were) formally offboarded or suspended

- ☐ Audit all third-party apps that have been installed by users on your domain. Examples of third-party apps include Chrome extensions, mobile apps, G Suite add-ons, and Office 365 add-ins. Determine the level of domain access requested by each application and revoke access/authorization to any apps if necessary

☐ Have a well-defined process for handling lost, stolen, or compromised devices

☐ Audit existing webhooks and scripts: Where do they live? Who's managing them? Do they have keys in them that shouldn't be there? How are the scripts hosted?

☐ Audit and remove any empty or unused groups, Slack channels, etc. to declutter your environment

☐ Consider creating a dedicated email group or Slack channel for IT that only contains relevant, actionable, and important alerts. Filtering out the noise helps reduce alert fatigue

As you work through that checklist, you can also start on a broader, more strategic SaaSOps action plan. This should be a long-term data protection plan, but it works in-line with the tactics listed above.

1. Identify your mission-critical applications and data.

First, start with a SaaS data inventory assessment exercise. Which applications should you prioritize? Which apps hold the most important data and are used by the most people? What kinds of information are in these apps? You can't protect what you don't know you have.

But not all your SaaS apps are business-critical. Some applications and data are more valuable than others. What information supports strategic business processes, objectives, and functions at your organization? Gather stakeholders to determine what SaaS data, if exposed or lost, could result in:

> Which apps hold the most important data and are used by the most people? What kinds of information are in these apps?

- Loss of revenue/customers
- Financial or regulatory fines/penalties
- Harm to the company's reputation, negative PR, or loss of consumer trust
- Loss of competitive advantage
- Business failure

2. Prioritize that data and create a data hierarchy.

What SaaS data needs very high protection and what doesn't? Prioritization is a cross-team effort. You'll need input from stakeholders (e.g., C-level executives, senior management, the board of directors, major shareholders and investors, the legal team, the data protection officer, the business units who own and interact with the data) to help classify and rank this data from most sensitive to least sensitive.

What data absolutely cannot end up in the wrong hands, be exposed, or be lost? What would you consider catastrophic? How would the data loss impact your organization's productivity, revenue, and security requirements (and for how long)?

How would the data loss impact your organization's productivity, revenue, and security requirements (and for how long)?

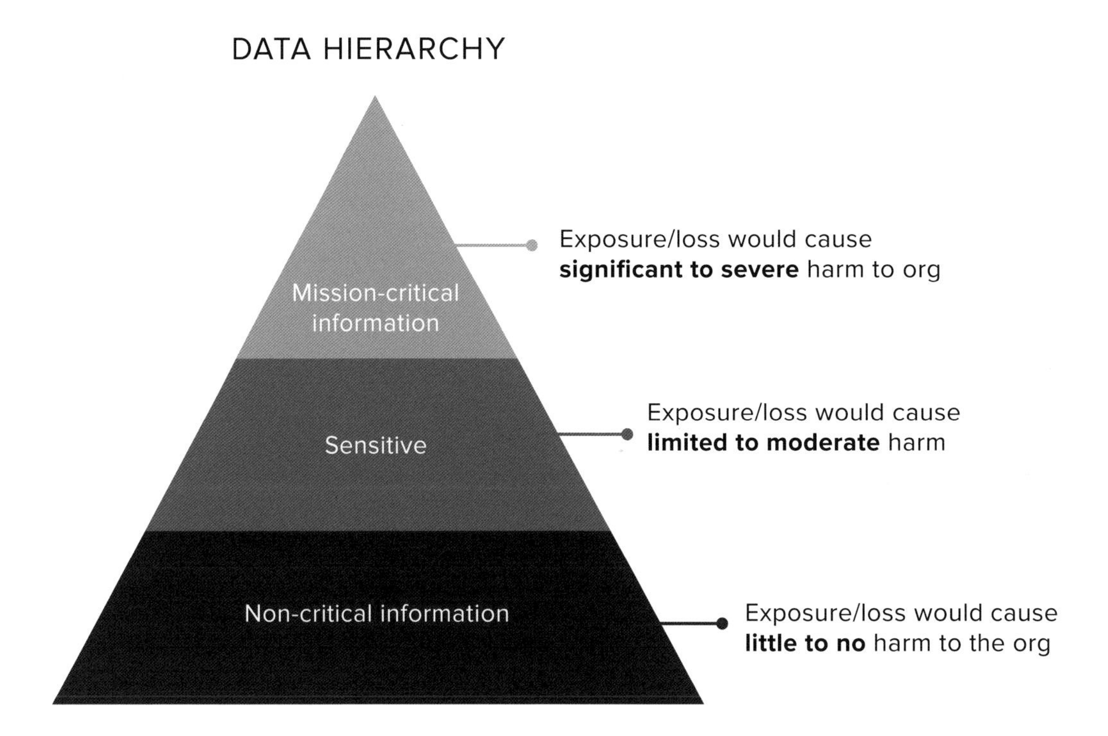

Figure 24: Sorting your data into a hierarchy will help prioritize what SaaS data needs high protection and what doesn't.

This step will likely include having a cross-functional, strategic conversation on your cyber risk appetite.

Every business is willing to take on a different amount of risk to achieve financial, reputational, and regulatory objectives. What's an acceptable risk tolerance for your organization? Furthermore, how much are you willing to invest to manage that risk? This will help prioritize the risks.

Where you fall on the SaaSOps Philosophy Spectrum is also reflective of your risk appetite. What controls do you have in place (or want to put in place) to address SaaS application risks, and what's an acceptable range of uncertainty related to those risks?

Your company will evolve and so will your cyber risk appetite. There will be shifts in operational risk and cyber risk. As such, determining your risk appetite shouldn't be a "one and done" exercise. It's a continual process and should be re-assessed often.

3. Create a policy framework that reflects your definition of trust.

Once you've prioritized your mission-critical SaaS apps and data, you're in a good place to lay the foundation for your policy framework.

Review your current security architecture and identify any gaps and sources of vulnerabilities. As you develop your new policy framework, think about how it can fill those gaps and protect your most mission-critical SaaS data. The questions below can provide guidance or help spark ideas as you shape your framework:

1. How are you defining trust?
2. Who are you considering a trusted vs. untrusted user?
3. Given the data you're protecting and your risk appetite, what interactions are you comfortable (and not comfortable) with your users having?
4. What types of user interaction do you need the most visibility into (but don't have today)?
5. Which user interactions are the riskiest for your organization?
6. What's your most mission-critical data and who should access be limited to?

7. Which dimensions are the most important for your organization?
8. Where is the comfortable equilibrium point for your organization—the point where the needs of the business and the needs of the user are balanced?
9. How would you have remediated the scenarios in the previous chapter?
10. How granular or sweeping do your policies need to be?
11. What kinds of policies would have the biggest impact on increasing operational efficiency?
12. Where is there room to automate operational processes?
13. What strategic work could your team take on if more of your operational processes were automated?
14. Is there currently a productivity/security tradeoff in your organization?
15. What are the biggest operational security risks and non-compliance areas?
16. What are your biggest organizational priorities right now?
17. How do you anticipate your operational controls and measures will change based on where your company will be in six months, one year, and five years?

Try using the template below as you're developing your policy framework. Substitute in different scenarios for your most mission-critical apps and data and think about what the best remediation path would be for your organization. This exercise can help illuminate which operational controls you need to put in place:

When ____________ **happens, and if** ____________
EVENT — OPTIONAL: ADDITIONAL CONDITIONS

is true, then ____________ .
REMEDIATION ACTION

Here are a few examples:

When a Box folder is shared publicly, and if it contains finance spreadsheets, then automatically remove all external collaborators, change the sharing setting to Private, and send the #security Slack channel a message.

When a user is leaving the company, then automatically reset user's password in Office 365, freeze the user in Salesforce, disable the user in Slack, remove the user from all Office 365 groups, revoke all of the user's third-party apps in Dropbox, remove all the user's devices in Office 365, remove the user and transfer their Dropbox files to their manager, create an email auto-responder in Office 365, send a reminder email in Office 365 to back up the user's data, wait 30 days, remove the Office 365 license, and delete the user.

When a user loses a device, then automatically reset their password in Okta, block their device in G Suite, delete their bypass codes in Duo, lock their device in AirWatch, and wipe their device in AirWatch.

When a contractor joins our Slack instance, and they have a 90-day contract, then wait for 83 days, send a Slack bot message to the contractor's manager asking if their Slack access needs to be extended or if IT can proceed with offboarding, wait for seven days, and disable contractor's account in Slack.

As you think through these questions, tie your responses back to your data hierarchy. With your definition of trust as your North Star, you can start establishing IT policies, guidelines, processes, and standards that align with company-wide organizational priorities.

Conclusion

SaaS applications are introducing new operational challenges and tensions in the digital workplace.

The key to securing your SaaS data is to secure user interactions, but interactions are spilling in all directions—between users, across apps, outside of your org—at a furious pace IT can't keep up with. It's also creating an enormous data sprawl that IT has little visibility into. Users are exposing data in new ways that IT isn't even aware of, creating new security issues. As a result, we're seeing the rise of a new category, SaaSOps, to address these emerging challenges.

At the same time, a painful tradeoff is rearing its head. Companies demands security and compliance, but users demand frictionless collaboration within their SaaS apps. Often these two forces are mutually exclusive.

With the tools on the market today, IT teams are at an impasse: sacrifice employee productivity or sacrifice security.

This double bind calls for a new way to manage and secure user interactions. It needs to be less rigid, less "all or nothing." That's because the concept of trust in the digital workplace is becoming nuanced, complex. It's not so black or white. It's less absolute than you might have thought.

As a result, how you manage and secure user interactions should be fluid, flexible—as lenient or strict as you need it to be. It should also be future-proof. Your company will undergo planned and unplanned changes and become subject to new regulations and standards. It may scale, shrink, pivot, shift, acquire, divest, and/or transform in the next few years. As your security philosophy and risk appetite evolve accordingly, so too should your SaaSOps strategy.

As a result, how you manage and secure user interactions should be fluid, flexible—as lenient or strict as you need it to be.

SaaSOps is the future of the digital workplace. Because the user is the new perimeter, IT teams must shift their attention to a new concept: user interactions. By thinking critically about user interactions now—what types of interactions are occurring, what trust looks like to you, the various dimensions of an interaction, where you fall on the SaaSOps Philosophy Spectrum—you can proactively and effectively start securing the data inside your apps.

ABOUT BetterCloud

BetterCloud is the first SaaS Operations Management platform, empowering IT and security teams to define, remediate, and enforce management and security policies for SaaS applications. Over 2,500 customers in 60+ countries rely on BetterCloud for continuous event monitoring, threat remediation, and fully automated policy enforcement. BetterCloud is located in New York City, San Francisco, and Atlanta, GA.

For more information, please visit **www.bettercloud.com**.

New to BetterCloud?

If you would like to learn more about BetterCloud, the first SaaS Operations Management platform, and work with a strategic technology advisor to address the needs of your organization, please visit **https://www.bettercloud.com/request-a-demo/**.

Existing BetterCloud customers:

If you would like to speak with your customer success manager about making the most of BetterCloud to manage and secure SaaS applications, please email **success@bettercloud.com.**

The IT Leader's Guide to SaaSOps (Volume 1): A Six-Part Framework for Managing Your SaaS Applications

If you enjoyed this book, you might enjoy Volume 1 of the series as well. Together, these books provide a complementary look at SaaSOps.

The IT Leader's Guide to SaaSOps (Volume 1) introduces the SaaS Application Management and Security Framework™, the first framework of its kind. This book was created by synthesizing insights from interviews, surveys, and conversations with thousands of IT professionals over the course of three years. By implementing this six-part framework, IT can gain an unprecedented level of control and clarity over its SaaS environment.

The IT Leader's Guide to SaaSOps (Volume 1): A Six-Part Framework for Managing Your SaaS Applications is available on Amazon.

By reading this book, you'll:

- Understand seven key operational challenges in SaaS environments
- Learn what an "enlightened" IT state looks like when the framework is executed
- Connect the framework to real-life news stories and data
- Visualize how all six elements of the framework build upon each other to create a high-performing IT environment
- Explore the range of solutions for executing the framework

"This book is amazing*—it will be incredibly useful for many IT and security professionals. I wish I'd had this book years ago. It makes me realize how much of a beginner I've been in terms of breaking down how to manage SaaS applications."*

— Andy Schwab, author of *Ultralight IT: A Guide for Smaller Organizations*

"This is a must read and should be mandatory for every modern IT professional. You should ***read this immediately to help audit, protect, and optimize your SaaS environment."***

— Ryan Donnon, IT and Data Manager at First Round Capital

"My company sells and implements over 50 SaaS applications. We do this for a living. I'm all too familiar with the challenges that companies face adopting SaaS, and this book nails them all. ***You are ahead of the game with this book."***

— Darryl Mitchell, Lead Solutions Engineer at NeoCloud

:party_cloud: